FOUNDATIONS FOR A NATIONAL BIOLOGICAL SURVEY

Edited by
Ke Chung Kim
and
Lloyd Knutson

Published by the
Association of Systematics Collections
In Cooperation With
Holcomb Research Institute
and
Illinois Natural History Survey

FOUNDATIONS FOR A NATIONAL BIOLOGICAL SURVEY

Published May 12, 1986

ISBN: 0-942924-13-4

Copies of **FOUNDATIONS FOR A NATIONAL BIOLOGICAL SURVEY** may be ordered from:

ASSOCIATION OF SYSTEMATICS COLLECTIONS
c/o Museum of Natural History
University of Kansas
Lawrence, KS 66045 U.S.A.
(913) 864-4867

CONTENTS

Foreword

I believe that history is on the side of systematics and with it the prospects for a national biological survey. To gain the support required it will be necessary to persuade the scientific community that systematists are gathering themselves for an effort that can carry along biology as a whole in some fundamental way, as well as one that conspicuously benefits society. The subject as a whole has not had much success of this kind in the recent past. Too often I have heard other biologists say, "When you give a taxonomist a grant, he goes off somewhere and writes a specialized monograph, and that's the end of it." And "Systematists haven't formulated a set of central questions in science that they alone are peculiarly qualified to answer."

Now the ambience is starting to change, for reasons having to do with two circumstances of historical necessity external to systematics. The first can be called the pluralization of biology. There are relatively few general principles and universal phenomena in biology at the molecular and cellular levels, and they have been pursued with such massive support and armies of scientists that while the field as a whole continues to advance impressively the number of significant discoveries per investigator per lifetime is in sharp decline. If I read the signs right, a new appreciation is growing for comparative studies, evolutionary reconstructions, and the study of particular groups of organisms in their own right. Put another way, the analysis of diversity is the new land of opportunity in biology: ten million or more species await, each with 10^6 to 10^9 bits of genetic information and a history ranging far back into geological time.

Meanwhile the practical need for knowledge of biological particularity grows more compelling each year. With the growing stress on the world environment, with each turn of the screw, the need for complete biological surveys becomes more obvious. For how will it be otherwise possible to monitor the decline of genetic diversity and shifts in ecosystems? How can biologists expect to pinpoint the species of most potential benefit to humanity during the tumultuous years ahead? Conducting research in applied biology without biological surveys is like trying to read an encyclopedia with a 200-word vocabulary.

The authors of *Foundations for a National Biological Survey* have expressed these exigencies very well. They provide authoritative accounts of the needs and promise of a national survey, as well as parallel efforts in other countries. The collection deserves to be read—and used—both by systematists and the other scientists and policy makers for whom diversity will be an increasingly important issue in the future.

Edward O. Wilson
Harvard University

Preface

A comprehensive knowledge of the species of plants, animals, and microorganisms in the U.S. and an understanding of their interrelationships have been fundamental goals of the biological sciences in the U.S. for well over 200 years. Much has been accomplished toward meeting those goals. Individual, regional, and serial taxonomic studies have been published, and major collections built. Federal, state, and special-purpose surveys have been conducted, and some continue to be active today. Yet we remain far from the needed, more complete state of knowledge of our flora and fauna. Over the past few years especially, a strong interest in a national biological survey has developed in the community of biologists.

There are diverse driving forces and interests on the part of both suppliers and users of a national biological survey. All of these interests seem to have a place in the development of a survey. They include interests and points of view of:

-individual researchers;
-professional disciplines, including the dozen or more national and regional societies that have recently passed resolutions in support of a biological survey and those yet to express their interests;
-public agencies (federal, state, and local); and
-universities, museums, and other private sector organizations.

Still other potential driving forces are, appropriately, waiting for the scientific community to engage in further peer review processes with regard to the nature, requirements, and benefits of a national biological survey before they support the endeavor.

Thus, the Association of Systematics Collections (ASC), as the national organization comprised of institutional members and concerned with systematics collections and progress in systematic biology, decided at its 1984 Annual Meeting to call for a major discussion on a national biological survey among the scientific public. At this meeting, the plight of the systematics community was discussed and priorities for the next decade suggested by K. C. Kim, H.-P. Schultze, C. Hubbs, P. F. Stevens, V. R. Ferris, and M. Kosztarab. Their concerns were echoed in the concept of a national biological survey for promoting systematic biology and a better understanding of the North American flora and fauna. This was in part an extension of steps taken toward a national biological survey by Michael Kosztarab through the ACS Council on Applied Systematics established in 1982 and chaired by K. C. Kim. As systematic entomologists—and thus having been confronted with the "worst case" situation, that is, the insects—we appreciated ASC's invitation to organize a broadscale national meeting on a biological survey.

In developing a program on this subject for the 1985 ASC Annual Meeting, we felt that the primary emphasis should be placed on the scientific and technical bases for a survey, the relationships of major scientific endeavors to a national biological survey, the linkages between a survey and the diverse user community, and the scope and benefits of a national biological survey. Although questions of funding, organizational basis, structure, and functioning are very important, it was felt that the scientific rationale should be pursued first. A national biological survey starts with specimens, with collections, and it is important to first consider these aspects, specifically and in detail. Collections are maintained largely to provide information for research and service purposes. We felt it important to concentrate on the data provided by collections—the kinds of data and how they are obtained, stored, managed, provided, and used. The authors were selected for their knowledge and experience in the specific areas and for the varying perspectives that they could bring to considerations of survey issues, not necessarily because they are or are not proponents of a national biological survey.

The purview of the 1985 ASC Symposium went far beyond the subject of collections. About 125 people attended the meeting. Many operational as well as technical points were raised in the formal and informal discussions: the relative degree of emphasis needed on tropical and North American survey work (and opportunities for synergism); potential competition of a national biological survey with existing programs (and opportunities for new funds); centralized vs. distributed models for a survey; extent of emphasis on ecological and environmental uses of survey data; neglect of certain groups, especially microorganisms; possible inherent values of a "national" effort; basic and applied emphases; geographic extent of a U.S. survey; and relationship of a national biological survey to national economic needs, to mention a few.

This book, primarily based on the proceedings of the 1985 ASC Symposium, is divided into six sections. In the first section, scientific bases for a national biological survey are discussed and the rationale and linkages of a survey described. The second section deals with the relevance and relationships of a national biological survey to agricultural, conservation, ecological, and environmental considerations. In the third section, the characteristics, structure, management, and dissemination of biological survey data are considered from several different perspectives. The fourth section deals with legislative and historical perspectives at federal, state, and private levels. In the fifth section, ambitious biological survey programs in Australia and Canada are described and related to a national biological survey for the U.S. The final section provides an overview and summary recommendations. We especially draw the reader's attention to these recommendations.

We are very grateful to the authors for their thoughtful and important contributions to the meeting and to this book. Special appreciation goes to Michael J. Bean for his important contribution, although he did not attend the meeting and belatedly participated in this effort.

We also thank Lorin I. Nevling and Stephen R. Edwards for inviting us to organize the meeting, John Richard Schrock, Stephen R. Edwards, and Sharon Mader of the ASC office for their excellent efforts in data entry and assisting in the editing of this book, and William L. Murphy of the Biosystematics and Beneficial Insects Institute, USDA, for his important editing and proofreading contributions.

Lloyd Knutson
Beltsville, MD

Ke Chung Kim
University Park, PA

February 1, 1986

Contributors

Michael J. Bean, Environmental Defense Fund, 1616 P Street NW, Suite 150, Washington, District of Columbia 20036

Peter B. Bridgewater, Director, Bureau of Flora and Fauna, G.P.O. Box 1383, Canberra, A.C.T. 2601, Australia

Barry Chernoff, Assistant Curator of Ichthyology, The Academy of Natural Sciences of Philadelphia, Nineteenth and the Parkway, Philadelphia, Pennsylvania 19103

Hugh V. Danks, Head, Biological Survey of Canada (Terrestrial Arthropods), Invertebrate Zoology Division, National Museum of Natural Sciences, Ottawa, Ontario, Canada K1A OM8

W. Donald Duckworth, Director, Bernice P. Bishop Museum, P.O. Box 19000-A, Honolulu, Hawaii 96817

Melvin Dyer, Senior Research Staff Member, Environmental Science Division, Building 1505, Oak Ridge National Laboratory, Oak Ridge, Tennessee 37830

Stephen R. Edwards, Executive Director, Association of Systematics Collections, c/o Museum of Natural History, University of Kansas, Lawrence, Kansas 66045

Michael Farrell, Director, Carbon Dioxide Information Center, Building 1505, Oak Ridge National Laboratory, Oak Ridge, Tennessee 37830

Allan Hirsch, Director, Office of Federal Activities, Environmental Protection Agency, 401 M Street, SW (A-104), Washington, District of Columbia 20460

Robert E. Jenkins, Jr., Vice President for Science and Natural Heritage Programs, The Nature Conservancy, 1800 North Kent Street, Arlington, Virginia 22200

Ronald L. Johnson, Senior Staff Officer, Animal and Plant Health Inspection Service—Plant Protection Quarantine, USDA, Room 609, Federal Center Building, Hyattsville, Maryland 20782

Maureen C. Kelly, Director for Document Analysis Division, Bioscience Information Services (BIOSIS), 2100 Arch Street, Philadelphia, Pennsylvania 19103

H. Edward Kennedy, President, Bioscience Information Services (BIOSIS), 2100 Arch Street, Philadelphia, Pennsylvania 19103

Ke Chung Kim, Professor of Entomology and Curator, Frost Entomological Museum, Department of Entomology, The Pennsylvania State University, University Park, Pennsylvania 16802

Waldemar Klassen, Director, Beltsville Area, Agricultural Research Service, USDA, Room 227, Building 003, Beltsville Agricultural Research Center—West, Beltsville, Maryland 20705

Lloyd Knutson, Director, Biosystematics and Beneficial Insects Institute, Agricultural Research Service, USDA, Room 1, Building 003, Beltsville Agricultural Research Center - West, Beltsville, Maryland 20705

Michael Kosztarab, Professor of Entomology, Department of Entomology, Virginia Polytechnic Institute and State University, Blacksburg, Virginia 24061

Orie Loucks, Director, Holcomb Research Institute, Butler University, 2600 Sunset Avenue, Indianapolis, Indiana 46208

Nancy Morin, Administrative Curator, Missouri Botanical Garden, P.O. Box 299, St. Louis, Missouri 63166

Lorin I. Nevling, Jr., Director, Field Museum of Natural History, Roosevelt Road at Lakeshore Drive, Chicago, Illinois 60605

Paul G. Risser, Chief, Illinois Natural History Survey, 176 Natural History Building, 607 East Peabody Drive, Champaign, Illinois 61820

Christine Schonewald-Cox, Research Scientist and Coordinator for NPS-UC Cooperative Studies Unit, National Park Service, USDI, Ecology Institute, Wikson Hall, University of California, Davis, California 95616

Stanwyn G. Shetler, Assistant Director for Programs and Curator, Department of Botany, National Museum of Natural History, Room W403 NHB, Washington, District of Columbia 20560

Wallace A. Steffan, Director, Idaho Museum of Natural History, Idaho State University, Campus Box 8096, Pocatello, Idaho 83209

Robert M. West, Director, Carnegie Museum of Natural History, 4400 Forbes Avenue, Pittsburgh, Pennsylvania 15213

Edward O. Wilson, Frank B. Baird, Jr. Professor of Science, Museum of Comparative Zoology, Harvard University, Cambridge, Massachusetts 02138

SECTION I.

INTRODUCTION

Scientific Bases for a National Biological Survey

Ke Chung Kim
The Pennsylvania State University

Lloyd Knutson
Biosystematics and
Beneficial Insects Institute

Abstract: The biota—fauna and flora—constitutes the living component of a natural ecosystem in which all organisms interact among themselves and also react and adjust in specific ways to their environments. The biota consists of unique, irreplacable resources, most of which are as yet untapped for human needs. It is fundamental to the utilization and management of the biota, the prevention of undue damage to natural ecosystems, and conservation of natural diversity that all of the component species are documented and the dynamics of the biota are understood. A national biological survey is the means to arrive at an understanding of these basic needs. The essence of a biological survey is an accumulation of systematic knowledge of the fauna and flora of a region, and thus biological surveys provide the essential database about living organisms for understanding ecosystem dynamics and conservation of living diversity. A biological survey is immediately linked to man's innate love of life and his own survival and is also directly relevant to a wide range of scientific disciplines and societal needs. Biological survey data provide in-depth databases for many fields and disciplines, such as agriculture, biogeography, ecology, entomology, parasitology, economic botany, tropical biology, and environmental sciences. In this chapter, such linkages are elaborated for ecology, parasitology, tropical research, and biotechnology. Other linkages between a national biological survey and various societal needs are also examined as an introduction to the following chapters. They include agriculture, environmental interests, germplasm resources, biological diversity, international programs, and state and federal programs.
Keywords: Biota, Flora, Fauna, Systematics, Ecology, Parasitology, Biotechnology, Agriculture, Environment, Diversity.

INTRODUCTION

The living species of plants, animals, and microorganisms must be documented and the processes fundamental to the maintenance of the biota must be understood

(Authorized for publication on September 26, 1985 as paper number 7266 in the Journal Series of the Pennsylvania Agricultural Experiment Station, University Park, Pennsylvania 16802, U.S.A.)

before we can fully utilize these organisms and knowledgeably monitor the health of our ecosystems. This need to enhance our understanding of the natural world is the primary basis for a biological survey, which is an organized effort to develop the systematic knowledge of the fauna and flora of a region. The biota—fauna and flora—includes all the organisms peculiar to a location, region, or specific environment and constitutes the living component of a natural ecosystem in which all organisms interact and in which each reacts and adjusts in specific ways to its environment. The structure and pattern of the biota are molded by such interactions.

Over evolutionary time, many new species have evolved and many others have perished through the dynamic process of interactions of biotic and physical factors. Extinction does not exclude the human species. During the last 100 years the human species has become the major unsettling force in the evolution of natural ecosystems and has contributed to the extinction of many organisms and to the imbalance of ecosystem dynamics. This process in turn threatens the sustained evolution of human civilization and causes concern for the very survival of our own species. Thus, prevention of undue damage to natural ecosystems and conservation of natural diversity should become biological imperatives for human survival. The first step in these efforts is an accurate assessment of the fauna and flora. In this paper we provide some opinions and conjectures in an attempt to clarify some of the scientific rationale and needs for a national biological survey.

THE BIOTA

The biota represents a contemporary diversity of living organisms, much of which is still poorly known. Since life began, many new species have evolved, and many others have become extinct. Forces of speciation and extinction continually changed the patterns of organismic diversity throughout the evolutionary history of the earth's ecosystem, and such evolutionary forces have molded the contempory biota.

The biota consists of unique, irreplaceable resources, most of which are untapped for human needs. The fauna and flora are sources of antibiotics, medicines, natural pesticides, industrial raw materials, food, fiber, fuels, ornamentals, and recreation. These resources must be protected from extinction, and biological surveys are essential to providing the basis of information for such protection.

In the early days of man as a hunter/gatherer, a mere 11,000 years ago, we were a symbiotic partner with other organisms in the natural ecosystems. These ecosystems represent the successional evolutionary changes of the more than two-billion-year history through which all extant species have evolved. During the last one million years, the human species emerged and sustained its evolution before industrial/military society took hold of human destiny. This suggests that the post-Pleistocene ecosystem has been favorable to human existence. Furthermore, the biotic structure of this ecosystem, for the last 15,000 years, has promoted the rapid but sustained evolution of modern man. If the human species is to survive and sustain its wellbeing during the next 10,000 years and beyond, we must conserve and properly manage what is left of the pre-industrial diversity of the earth's ecosystem.

Extinction is a natural process that has occurred throughout the evolutionary history of organisms. Mass extinctions of the Permian trilobites, Jurassic ammonites, and late Cretaceous dinosaurs demonstrate this evolutionary phenomenon. In fact, of the total number of species that have ever occurred on earth, more than 98 percent have become extinct (New, 1984). Extinction is caused by both biological factors and habitat alterations. Biological factors include competition, predation, parasitism, and diseases, whereas habitat alterations usually involve physical aspects such as geological change, climate, and catastrophe (Frankel and Soule, 1981). Extinction of a species results in the permanent loss of specific genetic diversity and may result in disturbance to the biological stability of the ecosystem.

In the last 2,000–3,000 years, the human species has emerged as a major element of the deterministic forces that have contributed to the extinction of many living organisms. As a result of the expansion of human habitation, agriculture, and commerce, and more recently, increases in industrial pollution, mankind has become the dominant biotic factor contributing to habitat alterations. As contemporary society expands, with an increasing population, human needs and uses of natural resources are exceeding what is available. To meet these societal demands, we abuse limited resources and inadvertently degrade and damage our environment. Accurate assessment of our natural resources and analyses of the rapidly changing human environment are the first steps toward a rational management of human ecosystems and sustained use of natural resources without eventually endangering human survival. Systematic knowledge of the fauna and flora is fundamental to this process, which will preserve biological values not only for the sustenance of the living diversity but also for the very survival of the human species.

SYSTEMATICS AND BIOTIC INVENTORY

Systematics as defined by Mayr (1969) is "the science dealing with the diversity of organisms" and is the most fundamental and inclusive discipline within the science of biology. Descriptions and classifications of living species provide the scientific basis upon which many other biological disciplines are developed, and taxonomic services provided by systematists make possible the research activities of scientists in other biological and environmental disciplines. These dual activities, which systematists have practiced for so long, are taken for granted by most other scientists and by the public. During the past several decades, that attitude has helped erode support for systematics; thus the livelihood of systematics as a science has drastically declined, as the Secretary of the Smithsonian Institution, Robert McC. Adams, correctly assessed (Holden, 1985). This trend has also resulted, in part, in the current critical shortage of taxonomic expertise (Kim, 1975a, b).

During the past 200 years, tremendous progress has been made in exploring the diversity of organisms. Approximately one million species of plants, animals, and microorganisms have been described, and more than 90 percent of the extant mammals and birds are relatively well documented. However, most other groups of living diversity remain poorly known; perhaps less than 10 percent of the extant species for many taxa such as the Insecta is known. Simpson (1952) estimated the total number of living species as two million, and Mayr (1969) considered it to range from 5 to 10 million species. Recently, Erwin (1982, 1983) estimated that insect species alone might number as high as 30 million when all undescribed insects in tropical rain forests, most of which have not yet been collected, are included. Considering the vast extent of living diversity, our knowledge of the earth's biota is at best meager. Many biological inferences based on the current database thus must be superficial. Systematists are confronted with an enormous task—to study and document this diversity before many species disappear and the stability of natural ecosystems is disturbed beyond repair.

Systematic biology, particularly alpha- and beta-taxonomy, as a science has declined in the U.S. during the past 30 years, resulting in a serious lack of the necessary taxonomic expertise and a shortage of basic data on the North American biota. Many universities have phased out systematic biology programs. This situation has in turn reduced employment opportunities for young systematists, particularly those in the field of taxonomy (Sabrosky, 1970; Kim, 1975a, b). A national biological survey could become a major vehicle to revive systematic biology and quickly improve the declining taxonomic expertise.

Biosystematics research is central to the conduct of a national biological survey. Without a proper emphasis on improving the extent of systematics research, a biological survey likely will fail. It is quite clear that the systematics community expects a national biological survey to provide increased support for systematics research and resources. However, systematists need to keep the broad objectives of a survey in mind, far beyond simply expecting it to support their own interests. Systematics and a national biological survey should not become what Pramer (1985) has called *terminal science*, which "... performs for its own selves or solely for the sake of scientists." Systematists and other biologists interested in a national biological survey must be responsive to the needs of the taxpayers.

BIOLOGICAL SURVEYS

A biological survey is an organized effort to study and document the fauna and flora. There are different concepts of a national biological survey. Whereas some consider the results of a survey to be primarily a series of identification manuals (which in fact are sorely needed), others place a high priority on an informational database, or network of databases, of value to user groups concerned with conservation, germplasm, pest management, etc. Still others emphasize basic taxonomic research. Obviously, these are interrelated and mutually beneficial results. But whatever results are desired, resources are needed. In all cases, it is the resources of collections and collection-associated data that would provide these final products and capabilities.

Biological survey activities by the Federal government began in 1885 with the formation of the Economic Ornithology Branch in the Division of Entomology, U.S. Department of Agriculture. One year later, this Branch was expanded to the Division of Economic Ornithology and Mammology. Ten years later, this became the Division of Biological Survey. Another decade later, in 1905, the Division became a full Bureau in the USDA. The Bureau was transferred to the Department of the Interior in 1939 and combined with the Bureau of Fisheries in 1940 to form the Fish and Wildlife Service (Gardner, 1984). Since then, most national biological survey activities have been gradually phased out, although certain projects, such as the Flora North America project, have contributed to survey objectives for certain periods of time.

Analysis of the different kinds of biological surveys being conducted throughout the world provides useful perspectives as to what direction might be taken in the development of a national survey in the U.S. Danks and Kosztarab (in press) provide an especially illuminating analysis in which existing biological surveys are reviewed by means of examples from regions that differ in faunal diversity, state of faunal knowledge, and available resources. They noted:

-"A 'complete' biological survey collects and preserves specimens, studies taxonomy, considers species inventory, distribution and faunal patterns within the region, and publishes the results of these investigations."

-"In addition, a survey is involved in ecological, experimental and evolutionary studies aimed at understanding as well as documenting the species composition and distribution, and in scientific coordination of studies, including advising government agencies or other authorities in matters related to the fauna and its value."

-"All of these roles seldom reside in one place or organization, but the broad concept of a biological survey requires that all of the roles coexist."

-"Such a survey can be coordinated from an overview of regional scientific problems and requirements, and through good communication among scientists. This can be assisted by an organization which provides liaison, and knows the whereabouts of resources for work on the fauna."

-"The chief value of the various types of biological surveys reviewed is that they can promote or extend basic taxonomic work on the fauna in three major ways: 1) By publishing broad works on the fauna (e.g., catalogues) and organizing faunal work into coherent series. 2) By coordinating work by various individuals and agencies to improve efficiency, and to broaden the use and integration of taxonomic and faunistic information; for example, by facilitating joint ventures in fieldwork, collecting, research, or synthesis of information. 3) By funding basic work in a cohesive way, or in a way that augments existing studies so as to remedy conspicuous deficiencies."

-"All of these roles stem from scientific requirements and thus must be organized through or by the scientific community and not in isolation."

-"Taxonomic surveys are necessary if organisms are to be identified and inventoried, but such an inventory is only one part of any attempt to characterize and understand regional faunas. What organisms do is a question to be asked in parallel with what they are by any biological survey. Recognition of this simple fact will help to provide broader support for biological surveys, both from individual scientists and from a variety of organizations concerned with entomological problems."

RELEVANCE OF BIOLOGICAL SURVEYS TO SCIENTIFIC ENDEAVOURS

A biological survey is not an extraneous, tangential, and inconsequential interest of relatively few systematists, but is directly, immediately, and importantly linked to man's love of life and his own survival. As Wilson (1984) perceptively stated, we must clearly recognize the environmental values of the living diversity and promote these values to maintain the survival of the human species.

A national biological survey is particularly important in this period when science is considered largely in the pursuit of imperatives of national security and economic gains. As science and engineering are increasingly embodied in the pursuit of these national imperatives, we should not lose sight of balancing perspectives for a world with unmet human needs and an increasingly fragile environment (Carey, 1985). This broader view is embodied in comments by Raman (1984) on science education:

> "Science as an intellectual enterprise has had little impact on the way people in general look at things. I contend that science should be taught not simply as a body of useful knowledge clothed in technical vocabulary but as a mode of inquiry into the nature of the perceived world, as an intellectual framework to guide us in the adoption of tentative interpretations of what is observed, and as a world view that is not ultimate truth but is applicable and acceptable only in the context of a given set of available facts. If that point of view is also encouraged in situations beyond technical problems, we may see a world where there is less dogmatism and greater mutual understanding. Science should be taught because of the value system it fosters, because of its criteria for the acceptance of points of view as valid propositions—beautiful and powerful theories. Science taught without reference to the scope and limits of human knowledge, without alluding to the collective nature of the enterprise, is incomplete."

We need to appreciate these kinds of concerns. A biological survey may indeed help this nation keep its fundamental and original philosophical and ethical aspirations alive and well, in addition to accomplishing specific program objectives of a national biological survey, such as production of detailed knowledge, manuals, identification of germplasm, and grist for the general research mill.

A national biological survey could serve as an all-encompassing rationale for study and preservation of flora and fauna, encompassing a broader diversity of disciplines than practically any other scientific enterprise. Biological survey data embody in-depth databases for many fields and disciplines, such as agriculture, biogeography, ecology, entomology, parasitology, economic botany, tropical biology, and environmental sciences. In the following sections, we will elaborate on the relevance of biological surveys to certain scientific endeavours.

A. Ecology and Parasitology

The many researchers in the broad spectrum of ecological research would be immediate users of biological survey data as gathered and arranged by systematists involved in a national biological survey. There is a perception that the relationships between these two major disciplines are not entirely what they should be, at either conceptual or operational levels. For example, in the announcement for the March 1985 symposium "Reflections on Ecology and Evolutionary Biology" at Arizona State University, it was noted that:

> "...in large part, ecologists and evolutionary biologists have neglected each other,

despite the early close association between their two fields. Why have ecologists sometimes, and sometimes not, found evolutionary thinking significant for their own research? And why have evolutionary biologists sometimes, and sometimes not, found ecological thinking significant for their research? Topics covered in the conference will include: the origins of ecology; the interrelationships between ecology, taxonomy, and biogeography; and the interrelationships between ecology and evolutionary biology" (Anonymous, 1984).

There is a special opportunity and need for a close relationship between long-term ecological research and a national biological survey. Long-term ecological research generates data and gathers material of direct importance to a survey and at the same time is in great need of the taxonomic information that can be derived from such data and collections. Especially, predictive taxonomic classifications can provide useful insight into long-term ecological research. Such linkages could include not only the Long-Term Ecological Research Program (LTER) at 12 sites with National Science Foundation (NSF) support, but the several programs supported by state and private groups. The faunistic and floristic information needs of NSF-supported LTER programs, needs ranging from data on nematodes in grasslands to data on fingernail clams in the Mississippi, have recently been analyzed (Lattin and Stanton, in prep.). For example, it is important for those working on energy flow networks to know if the nodes represent one or several species. The broad range of needs and opportunities relating to long-term environmental research and development, including biological monitoring and survey, was discussed in detail in a draft report of a conference held by the Council of Environmental Quality (CEQ, 1985). Among the 13 principal issues identified as warranting particular emphasis were:

- Improving the quality and cost-effectiveness of physical, chemical, and biological monitoring programs to test scientific hypotheses about how environmental systems operate and interact.
- Expanded collection of 10-year observations of the background physical, chemical, and ecological variations in fresh waters, oceans, and the atmosphere.
- Development of biological inventories and baseline studies of ecosystem structures, functioning, and linkages.

In his presidential address to the 58th Annual Meeting of the American Society of Parasitologists, Heyneman (1984) emphasized the urgent need for new techniques for initial surveys of large areas, such as field-adapted Telestar and Landsat infrared scans for remote sensing, citing the Global Environment Monitoring System (Gwynne, 1982). He reiterated the point stated in the *Science* editorial by Kosztarab (1984)—that less than one-third of the organisms and their developmental stages have been studied in the U.S. Heyneman (1984) aptly described the need for taxonomic information relevant to parasitology:

"Equally important would be simplified computer access to the vast biological systematics collections available only in the world's major museums—millions of man-years of research now only accessible to and interpretable by rapidly diminishing numbers of taxonomic specialists. Far wider access to these data, keyed to geographic, geologic, biologic, and social factors, is essential for zoonotic researchers to identify faunal elements and to predict possible epidemics among new settlers, or anticipate outbreaks to be expected with enforced migrations through specified ecologic regions. The enormous potential of computer-based technology for storage, integration, and worldwide access to biomedical information underscores the wretched state of funding of systematics studies, the nonavailability of taxonomists, and the lack of proper training for them—at a time when this knowledge and the specialists to continue to provide it have never been so needed, so underrated, and so unrewarded."

B. Tropical Research

Tropical habitats provide special opportunites for many kinds of interesting and important research concerning, for example, diversity, relict floras and faunas, and germplasm and vanishing habitats. Several important tropical habitats lie within continental North America (e.g., southernmost Florida) and Hawaii, and these are among the areas most disturbed by humans. A national biological survey certainly encompasses these tropical habitats.

Is there intrinsic conflict or need for competition for support between surveys in the tropics and a biological survey of the U.S.? Because the "biological exploration of the planet Earth" is one of the ultimate goals of biology, and because the tropical biota has a certian priority in this exploration, considerable attention needs to be paid to the relationships between research in the tropics and a U.S. national biological survey. Just as tropical-temperate studies have been synergistic in many ways in the past, they can be mutually beneficial in further, in some cases unpredictable, ways. Taxonomic and phylogenetic studies of cosmopolitan and widely-distributed taxa make it obligatory to know both tropical and temperate faunas and floras.

C. Biotechnology

Increased research activities in recent years in molecular biology and its allied disciplines have generated an explosion of knowledge in fields such as systematics, developmental biology, genetics, biochemistry, neuroscience, and others. To an ever increasing extent, this knowledge has been directly applied to the production of useful products (biotechnology). It makes sense now for every discipline not only to examine how it can use molecular biology to advance knowledge but also to understand how it can impact on biotechnology.

Biotechnology has been variously defined, from the "old biotechnology" definitions used in Europe and Japan—"the application of engineering and technological principles to the life sciences"—to the "new biotechnology" definitions, predominant in the U.S., which emphasize recombinant DNA techniques. The matter of definition is important because a biological survey enterprise would have a broader or narrower relationship with biotechnology, depending upon whether we are referring to the broad or narrow definition. A recent article in *BioScience* reviews some of the definitions and outlines some theoretical and empirical relations between the "new" (rDNA-based) biotechnology and molecular biology (Markle and Robin, 1985). The definition originally adopted by the Office of Technology Assessment (which also was adopted by other organizations and, for a period, by NSF) is well-balanced and offers room for involvement of some research related to a national biological survey. That is, "Biotechnology... includes any technique that uses living organisms (or parts of organisms) to make or modify products, to improve plants or animals, or to develop microorganisms for specific uses" (Office of Technology Assessment, 1984).

Recently, NSF has established an Office of Biotechnology Coordination, partly in response to a 1984 request by the Office of Science and Technology Policy's Cabinet Council Working Group on Biotechnology (Wortman, 1985). The Group called for NSF "...to examine the potential effects of environmentally related basic research in biotechnology...," building "...on the strong ecology and ecosystem research program currently operated by NSF." The NSF office has established an accounting system for NSF-supported research relating to biotechnology and assisting in the review of research proposals, especially risk assessment. The first head of the new office, Dr. Robert Rabin, had noted, "We [NSF] can't use the definition proposed by the Office of Technology Assessment in 1984. It's too broad." (Wortman, 1985). However, concern about areas of research related to biotechnology (in the context of NSF expenditures on biotechnology) was indicated by Rabin's statement, "We must seek advice from many people and determine a definition of *relatedness to biotechnology* as well" (italics original). Although proposals to NSF are now being coded and analyzed from the viewpoint of relatedness of 20 fields, "systematics" is not one of these, but "environmental biology" and "special resources" are.

Many biotechnology methodologies are being used in systematics and evolutionary biology research. The development of genetic libraries of diverse organisms and the improvement of methodologies are complementary interactions. Indeed, evolutionary questions of systematics are a major driving force in the biotechnology/molecular biology area.

It is disappointing, however, that systematics of higher organisms and related disciplines, which provide the basis for a national biological survey, have so far not made strong conceptual or practical linkages with biotechnology programs, especially those emphasizing the "new biotechnology." Some research planners in fact have recognized the relationship of biosystematics and related research to biotechnology. For example, in an analysis of biotechnology-related research in the Beltsville Area of the Agricultural Research Service, "Modern Methods for Taxonomy and Germplasm Evaluation" was identified as one of six areas of emphasis, and the six major specimen collections were specifically highlighted as special resources for biotechnology (Purchase, 1984). Modern biochemical and genetic methodologies as used in systematics research can provide the kinds of data useful to biotechnology. Those research environments bear watching for future opportunities.

Further opportunities for a useful relationship between biosystematics and biotechnology will emerge via ecological research. Knowledge of the identity, distribution, and biosystematics of our endemic organisms, as well as those of other parts of the world, is a key resource for biotechnological research. Biotechnologists should recognize that they are now making use of the extensive knowledge that conventional systematics research and survey work has stored up in the past. How long will these information resources be sufficient? How much more efficiently could the biotechnologists work if they were dealing with a more complete resource of biosystematic information? If the predictive values of classifications were improved, along with the development and implementation of computerized networks of databases to provide the information, how valuable would these be to the biotechnology industry?

Natural products research may offer another linkage. For example, take the case of natural plant chemicals as sources of industrial and medicinal materials. Balandrin et al. (1985) pointed out that only 5 to 15 percent of the 250,000 to 750,000 existing species of higher plants have been surveyed for biologically active natural products, and, in general, the investigated species have been examined for only one or a few types of activity. With regard to a biological survey, two other major points emerge from their paper. First, the analytical power of the diverse methodologies available today is truly remarkable. The "needle in the haystack," the rare occurrence of a highly desirable commercial genotype, can in fact be located. Second, the very methodologies of recombinant DNA technology alleviate the need for large amounts of rare living material; this is what brings to the forefront the combined importance of the taxonomy of obscure organisms and the predictive power of classifications. For example, large-scale production of a rare but genetically useful plant would not be required for investigation and production of its product.

In the process of providing such "up-front" support for biotechnology, systematics and ecology are also providing important direction and support for the *results* of biotechnology. Forcella (1984) recently provided some forward-looking thoughts and instructive examples with regard to these points. He commented, "Ecologists are the people most fit to develop the conceptual directions of biotechnology. We are the ones who should have the best ideas as to what successful plants and animals should look like and how they should behave, both individually and collectively." He asked if ecologists should "... take the forefront in biotechnology, and provide the rationale for choosing species, traits, and processes to be engineered? I suspect this latter approach will be more profitable for the world at large as well as for ourselves." There is, or should be, direct and intimate interactions between systematists and ecologists, as well as geneticists, in the identification and selection of species and traits for biotechnological application, which are much broader than strict concern about environmental safety. A biological survey is directly related to knowledge about such species and traits, and knowledge of many processes and ecological principles are dependent on the kinds of resources that can be provided by a biological survey.

LINKAGES TO NATIONAL NEEDS

One of the primary objectives of the ASC symposium on a national biological survey was to examine linkages between a biological survey and other groups, interests, and areas of activity. Several chapters in this book will explore these aspects in more detail. Although there may be some overlap in some of the subjects considered, we find significant value in coming at the issues from different viewpoints.

With the unifying target of understanding the flora and fauna of the U.S., it is possible to include many other specialized areas of interests. If we really believe that elucidation of the plants and animals of the U.S. is a valid goal, then long-term plans are needed. Some people fear that short-term projects, especially those with an economic emphasis, detract from the main course. We do not think so.

A. Agriculture

A number of areas in agriculture rely heavily on biological survey data. Among these areas are plant protection and quarantine (Shannon, 1983), biological control (Sabrosky, 1955), germplasm research (Schonewald-Cox, 1985), and gene and biotype banking.

The broad range of phenomena applicable to the study of colonizing species is an area of research that needs further development along conceptual lines. In agriculture, there is special interest in this area from the point of view of both introduced biological control agents and immigrant pests. But it is a fact that there is not even a simple listing of the immigrant arthropod species in the U.S. In the USDA's Biosystematics and Beneficial Insects Institute, such a computerized database is being developed—nearly 2,000 immigrant species are known to date, but the total will probably be more like 6,000. The ABRS "Bioclimate Prediction System" mentioned by Peter Bridgewater in this volume is another example of the application of survey data.

Biological control by beneficial insects began in this country in 1884 with the introduction of the Australian vedalia beetle into San Diego for control of the Australian cottonycushion scale. The California citrus industry was saved from virtual elimination for the cost of $2,000 for an introduction trial. Since then, over 600 species of parasites, predators, and phytophagous insects and mites have been released by federal and state units to control a number of pest insects and weeds. Nearly half, about 200 parasites and predators and 25 weed-feeding arthropod species, are known to be established. Over 35 species of pests are being substantially or completely controlled by introduced beneficials in at least some portions of their range in the U.S. These figures are probably conservative. Until 1982, when the USDA's ARS Biological Control Documentation Center and National Voucher Specimen Collection were established in the USDA's Beneficial Insects Laboratory at Beltsville, Maryland, there was no procedure for systematically capturing information on releases and establishments. That deficiency works against support for biological control, the most economical, environmentally sound, and energy-conservative method of pest control. That deficiency also contributes to a certain lack of rigor in the planning and conduct of biological control research. Clearly, there is an important linkage between biological control and a national biological survey.

B. Environmental Interests

Certain kinds of data derived from a national biological survey need to be considered from the point of view of those who must meet the requirements of the National Environmental Policy Act of 1969. The environmental protection and conservation communities are well aware of this linkage.

In reviewing the Conservation Foundation's 1984 report "State of the Environment: An Assessment at Mid-Decade," Pimentel (1985) makes several cogent points:

"An encouraging assessment [in this book] of wildlife focuses primarily on birds, fishes, and mammals. Although much progress in protecting these animals has been made in the past 20 years, birds and mammals constitute less than 1% of the total biomass in our terrestrial ecosystem. About 200,000 other species [in the U.S.] dominate the system, both in numbers and biomass. Many of these organisms degrade and recycle wastes; some microbes fix nitrogen; bees pollinate some crops. Many of these organisms are seriously affected by chemical pollution. As too often happens, the report overlooks the importance of small organisms vital to the ecosystem."

C. Germplasm Resources and Biological Diversity

Stemming from the 1981 Strategy Conference on Biological Diversity, other conferences, and specifically from the International Environment Protection Act of 1983, the U.S. Congress asked the U.S. Agency for International Development (AID) to develop a strategy on the protection and conservation of biological diversity in developing countries. The report (Interagency Task Force, 1985):

"...describes the basis for concern over the loss of biological and genetic diversity in developing countries ...outlines current activities and programs undertaken by Federal agencies to deal with natural resources conservation", and "...presents a U.S. strategy for building on those activities and programs to assist developing countries."

The report, which was reviewed by Tangley (1985), is of interest to a U.S. national biological survey from several points of view. Because the strategy includes the identification, taxonomy, and cataloging of animal and plant species, especially in tropical environments, it should provide yet another opportunity for the positive, synergistic effects of collections-based work in the U.S. and elsewhere. Through such programs, increased data collection and biotic studies in developing countries likely would involve systematists and other specialists in the United States. U.S. national biological survey activities and the AID-initiated studies in developing countries would contribute to our knowledge of the world biota, and thus these studies would complement each other in a synergistic fashion. The recent publication of "Resource Inventory and Baseline Study Methods for Developing Countries" (Conant et al., 1984) by the American Association for the Advancement of Science also indicates the increasing international interest in biological diversity.

D. International Programs

There are major international arenas in which a U.S. national biological survey obviously should be involved. One is the "International Geosphere-Biosphere Program—A Study of Global Change", to be coordinated by the International Council of Scientific Unions. The preface to the report of a 1983 National Research Council Workshop gives an impression of this developing Program (NRC, 1983):

-"If, however, we could launch a cooperative interdisciplinary program in the earth sciences, on an international scale, we might hope to take a major step toward revealing the physical, chemical, and biological workings of the Sun-Earth system and the mysteries of the origins and survival of life in the biosphere. The concept of an International Geosphere-Biosphere Program (IGBP), as outlined in this report, calls for this sort of bold, 'holistic' venture in organized research—the study of whole systems of interdisciplinary science in an effort to understand global changes in the terrestrial environment and its living systems. The National Research Council IGBP Workshop at Woods Hole in July 1983 considered the major problems for research in five areas that might naturally be coordinated in such a program: the atmosphere, oceans, lithosphere, biosphere, and solar-terrestrial system."

-"If we believed that the Earth was a constant system in which the atmosphere, biosphere, oceans, and lithosphere were unconnected parts, then the traditional scientific fields that study these areas could all proceed at their own pace treating each other's findings as fixed boundary conditions."

-"A major challenge to an IGBP will be that of understanding the causes and effects of climate change."

-"Advantages equally great have come in the past 20 years from orbital sensing of the Earth's biota. Much of the motivation has been to develop means of assessing conditions for the world production of food, fiber, and fuel from renewable biological resources. The focus has been on agricultural crops, forests, and rangeland of economic importance, and the principal tool has been infrared mapping. In recent years, microwave techniques have come to the fore, offering added capability in sensing through overcast and penetrating more deeply into canopies of vegetation."

Biosphere aspects of the International Geosphere-Biosphere Program are, in turn, related to four other ongoing international programs:

1. "Man and the Biosphere"
2. "Analyzing Biospheric Change"
3. "A Decade of the Tropics"
4. "Global Environmental Monitoring System"

Other international, multidisciplinary research programs are the Global Habitability Program proposed by the United States at the United Nations in 1982 and the program for Climatic, Biotic, and Human Interactions in the Humid Tropics. The advantages and disadvantages of such large national and international coordinated programs, and previous programs such as the International Biological Program, need to be evaluated in relation to a national biological survey.

Edelson (1985) commented on the essence of this relationship in a *Science* editorial entitled "Mission to Planet Earth":

"In some ways we know more about our neighboring planets than we do about the earth. For decades scientists have peered at Venus and Mars through telescopes, and in the last decade they have had radar images of the planet surfaces made from the earth and from orbiting satellites. They have probed the atmospheres of these planets and measured and sampled their surfaces with instruments of the space age. Of course, through the centuries we have accumulated a mountain of detailed data points and much phenomenological knowledge about the earth and the constituents of its geosphere and biosphere. However, we lack synoptic, systematic, and temporal knowledge of our own planet and an understanding of the mechanisms underlying the global processes that affect it."

E. State and Federal Mission Agencies

One might expect some difficulties in relating an extensive, broad-framework enterprise such as a national biological survey to the diverse state and federal mission agencies. We do not believe that is true. The USDA's Agricultural Research Service, for example, is a highly mission-oriented agency. One of ARS's basic operational tenents is that it is a problem-solving organization. However, one of the three ARS key strategies is to "Maintain emphasis on mission-oriented, fundamental, long-range, high-risk research"; this is a strategy relatively consonant with a national biological survey. And although ARS supports systematics research in the areas of arthropods, nematodes, animal parasites, fungi, and vascular plants, employing about 40 research taxonomists, the programs of those laboratories are specifically directed to meet the needs of the agency. It would be fair to expect that mission agencies will carefully examine a national biological survey and interact with those aspects that are rather closely related to their missions.

CONCLUSIONS

The essence of a biological survey is the accumulation of systematic knowledge of the fauna and flora of a region. Biological surveys provide this essential database of knowledge about living organisms for systematics and other related scientific endeavours. Knowledge of the biota provides not only the material basis for a better understanding of organisms but also embodies environmental ethics and biological values which help us to appreciate the living diversity and our own existence in the deteriorating human ecosystem.

The biota in a natural ecosystem is an evolutionary manifestation. To reach the post-Pleistocene ecosystem in which the human species emerged and successfully evolved, all the organisms interacted among themselves and with their environments, and such interactions molded the present structure and pattern of the fauna and flora. Extinction as a natural process has occurred throughout the evolutionary history of organisms. The human species is, unfortunately, not excluded from this process. The extant species of the earth's biota are, in a sense, successful survivors of continuous, intense evolutionary battles of more than two billion years. To understand better the fauna and flora and to prevent living species from unnecessary extinction are biological imperatives for human survival.

In relatively recent times, the human species has emerged as a major force contributing to the extinction of many organisms and the instability of natural ecosystems. Modern agriculture, industry, and human habitation have greatly rearranged the pattern and adversely affected large sections of natural ecosystems through which species have disappeared (Marine, 1969). In the 1960's, Americans, their thinking focused by the writings of the environmental crusader Rachel Carson, reacted against the use of DDT and other persistent pesticides. This movement led to the passage of the National Environment Policy Act and promoted an environmental conscience among the American public. However, toward the year 2000 we must even more clearly recognize the biological values of living diversity in the pursuit of imperatives for human survival.

The essence of a biological survey is recognized by, and identified with, society in the form of organizations and groups that touch the lives of people in various ways, i.e. the entire range of organism-oriented organizations (Audubon Societies, The Nature Conservancy, World Wildlife Fund, Small Farms, Clean Air, Clean Water, etc.), to various political action groups. The potential extent of relationships between a national biological survey and such groups is enormous. Consider, for example, only the area of conservation—the National Wildlife Federation's 1984 Conservation Directory (National Wildlife Federation, 1985) lists more than 2,500 organizations and 13,000 individuals in the U.S. and 113 other countries.

Knowledge of the North American biota is necessary because:

1) We are a member of the biota, and we will be affected by changes in the natural ecosystem;

2) Biospheric changes due to expanding human populations and man-made pollutants are threatening the size of the biota and its genetic structure;

3) Systematic knowledge of the fauna and flora is fundamental to the study of the evolutionary history of organisms and man's place in the ecosystem;

4) Accurate inventory of today's fauna and flora is fundamental to the monitoring of changes in human ecosystems, forged by integration of man-biota-total environment, which provides for our very basic survival;

5) The sciences of living diversity provide the means to sustain basic biological, environmental, and human values.

RECOMMENDATIONS

1. An assessment of the current state of knowledge of the biota in the United States should be made at the initiation of the survey.
2. Further planning for a national biological survey should emphasize analysis of specific conceptual and operational linkages between a biological survey and specific institutions, disciplines, projects, and areas of research.
3. Systematists, ecologists, and applied scientists in industry could productively and jointly analyze the relationships of a national biological survey to the broad area of biotechnology.
4. Research leaders and planners with specific research-area expertise, as well as representatives of the private sector, should be directly involved in planning a biological survey.
5. Other national and international meetings now being planned will deal more extensively with aspects such as funding, organizational structure, and management. At the same time, it should be kept in mind that an overriding requirement is that plans for implementation need to be considered in the context of the nature and use of the information.

ACKNOWLEDGEMENTS

We thank F. Forcella, North Central Soil Conservation Research Laboratory, Agricultural Research Service, U.S. Department of Agriculture (ARS-USDA), Morris, Minnesota; R. W. Hodges, M. D. Huettell, W. L. Murphy, A. L. Norrbom, and H. G. Purchase, Beltsville Agricultural Research Center, ARS, USDA; H. E. Waterworth, National Program Staff, ARS, USDA; C. W. Pitts and J. Schultz,

The Pennsylvania State University; S. G. Shetler, Smithsonian Institution; J. L. Brooks and associates, Biotic Systems and Resources, National Science Foundation; and S. R. Edwards, Association of Systematics Collections, for their review of the manuscript.

LITERATURE CITED

Anonymous. 1984. *Announcing reflections on ecology and evolutionary biology.* Arizona State University, Tempe, Arizona. March 1–2, 1985. 2 p., unpubl.

Balandrin, M. F., J. A. Klocke, E. S. Wurtele, & W. H. Bolinger. 1985. Natural plant chemicals: sources of industrial and medicinal materials. *Science* 228: 1154–1160.

Carey, W. 1985. Editorial: Science: matters of scale and purpose. *Science* 228: 7.

Conant, F., P. Rogers, M. Baumgardner, C. McKell, R. Dasmann, & P. Reining (eds.). 1984. *Resource inventory and baseline study methods for developing countries.* Amer. Assoc. Adv. Sci., Washington, D.C. 565 p.

Conservation Foundation (The). 1984. *State of the environment: an assessment at mid-Decade.* Washington, D.C. 586 p.

Council on Environmental Quality. 1985. *Report on long-term environmental research and development.* U.S. Government Printing Office. 11 p. + 5 append.

Danks, H. V. & M. Kosztarab. In press. Biological surveys. *In*: L. Knutson, K. M. Harris, & I. M. Smith (eds.) *Biosystematic services in entomology.* Proceedings of a Symposium held at the XVIIth International Congress of Entomology, Hamburg, Federal Republic of Germany, 1984.

Edelson, B. I. 1985. Letter to the editor. Mission to planet earth. *Science* 227: 367.

Erwin, T. L. 1982. Tropical forests: their richness in Coleoptera and other arthropod species. *Coleopt. Bull.* 36(1): 74–75.

Erwin, T. L. 1983. Tropical forest canopies: the last biotic frontier. *Bull. Entomol. Soc. Amer.* 29(1): 14–19.

Forcella, F. 1984. Commentary. Ecological biotechnology. *Bull. Ecol. Soc. Amer.* 65(4): 434–436.

Frankel, O. H. & M. E. Soule. 1981. *Conservation and evolution.* Cambridge University Press, Cambridge.

Gardner, A. L. 1984. Letter to the editor: biological survey. *Science* 224: 1384.

Gwynne, M. D. 1982. The global environment monitoring system of UNEP. *Environmental Conserv.* 9: 35–41.

Heyneman, D. 1984. Presidential address. Development and disease: A dual dilemma. *J. Parasitol.* 70(1): 3–17.

Holden, C. 1985. New directions for the Smithsonian. *Science* 228: 1512–1513.

Interagency Task Force. 1985. *U.S. strategy on the conservation of biological diversity. An interagency task force report to Congress.* U.S. Agency for International Development. Washington, DC. x + 54 p.

Kim, K. C. 1975a. Systematics and systematics collections: Introduction. *Bull. Entomol. Soc. Amer.* 21(2): 89–91.

Kim, K. C.. 1975b. A concluding remark. *Bull. Entomol. Soc. Amer.* 21(2): 98–100.

Kosztarab, M. 1984. Editorial: A biological survey of the United States. *Science* 223: 443.

Lattin, J. D. & N. L. Stanton (in prep.). *Report on workshop for ecologists and systematists on priorities for collaborative work on soil organisms.* Oregon State University, Corvallis, Oregon. May 20, 1985.

Marine, G. 1969. *America the raped.* Discus Books, New York.

Markle, G. E. & S. S. Robin. 1985. Biotechnology and the social reconstruction of molecular biology. *Bioscience* 35(4): 220–225.

Mayr, E. 1969. *Principles of systematic zoology.* McGraw-Hill, New York.

National Research Council (NRC), Commission on Physical Sciences, Mathematics, and Resources. 1983. *Toward an international geosphere-biosphere program. A study of global change.* Report of a National Research Council Workshop, Woods Hole, Massachusetts, July 25–29, 1983. National Academy Press, Washington, DC. xiii + 81 p.

National Wildlife Federation. 1985. *Conservation directory 1984.* 29th edition. Washington, DC.

New, T. R. 1984. *Insect conservation: an Australian perspective.* Series Entomologica (Editor K. A. Spencer), Vol. 32. Dr. W. Junk Publishers, Dordrecht, Netherlands. 184 p.

Office of Technology Assessment. 1984. *Commercial biotechnology: international analysis.* U.S. Government Printing Office, Washington, DC.

Pimentel, D. 1985. Book review: Update on the environment. State of the environment: an assessment at mid-Decade. The Conservation Foundation, Washington, DC. 1984. *Bioscience* 35(3): 198.

Pramer, D. 1985. Opinion: Terminal science. *Bioscience* 35(3): 141.

Purchase, H. G. 1984. *Agriculture biotechnologies. Strong acceleration of research programs at Beltsville.* U.S. Dept. Agric., Agric. Res. Serv. 28 p.

Raman, V. V. 1984. Commentary. Why it's so important that our students learn more about science. *J. Wash. Acad. Sci.* 74(3): i–iii.

Sabrosky, C. W. 1955. The interrelations of biological control and taxonomy. *J. Econ. Entomol.* 48: 710–714.

Sabrosky, C. W.. 1970. Quo vadis, taxonomy? *Bull. Entomol. Soc. Amer.* 16(1): 3–7.

Schonewald-Cox, C. M. 1985. Diversity, germplasm, and natural resources. *In*: K. C. Kim & L. Knutson (eds.) *Foundations for a national biological survey*, Association of Systematics Collections.

Shannon, M. J. 1983. Systematics: A basis for effective regulatory activities. *Bull. Entomol. Soc. Amer.* 29(3): 47–49.

Simpson, G. G. 1952. How many species? *Evolution* 6: 342.

Tangley, L. 1985. A new plan to conserve the earth's biota. *Bioscience* 35(6): 334–341.

Wilson, E. O. 1984. *Biophilia.* Harvard University Press, Cambridge and London. 157 p.

Wortman, J. 1985. NSF sets up Office of Biotechnology Coordination. *Bioscience* 35(6): 340–341.

SECTION II.

ECOLOGICAL AND ENVIRONMENTAL CONSIDERATIONS

Prefatory Comments: Some of the Activities Leading to this Symposium

Michael Kosztarab
Virginia Polytechnic Institute
and State University

The following introductory comments are intended to provide background information, especially on the activities of a planning committee for a national biological survey that helped to lead to the Association of Systematics Collections' (ASC) national meeting entitled "Community Hearings on the National Biological Survey". Some of the practical applications anticipated from a national biological survey are also discussed.

Earlier attempts to initiate similar surveys included efforts by government agencies such as the U.S. Department of Agriculture (Cameron, 1929) and Fish and Wildlife Service (Gabrielson, 1940); by scientific organizations, such as the Flora North America project sponsored by the American Society of Plant Taxonomists; and by individual scientists, such as the Faunal Survey of North America by M. D. F. Udvardy (personal communication, 1966) and the Insects of North America project by Kosztarab (1975). Unfortunately, none of these attempts succeeded in initiating the much needed comprehensive biological survey. In the United States, biologists have described only one-third of the living organisms and their developmental stages (Kosztarab, 1984a). One scientist compared this deficiency to a factory that is run without an inventory. Most developed nations have already prepared or are developing such inventories. Because of the rapid decline of our biota, the proposed biological survey is a national imperative and should be an urgent interdisciplinary program supported by the federal government, as it is in Canada (Danks, 1978), Australia (Bridgewater, 1984), and other countries.

The need for a biological survey, especially on insects and related poorly known taxa, was recognized by the Standing Committee on Systematics Resources (SCSR) of the Entomological Society of America (ESA). The Committee has been chaired through the years by a number of systematists, including K. C. Kim, Lloyd Knutson, and myself. When my appointment as chairman terminated in 1983, I was asked by the new SCSR chairman, Paul M. Marsh, and encouraged by the ESA leadership to organize and chair a subcommittee within SCSR to promote the initiation of an insect faunal survey, later named Insect Fauna of North America

(IFNA) project. The work of our subcommittee led to the realization that there are many other neglected taxa besides insects, and such a survey effort should be expanded to the entire biota. I was encouraged to do so by a number of colleagues representing a variety of biological disciplines, who, after reading my editorial in *Science* (Feb. 3, 1984) joined in my efforts and offered to serve on a broadened committee promoting a national survey of the entire biota. This led to the formation of a planning committee for the national biological survey on March 7, 1984, at the National Museum of Natural History. At this meeting Charles Cushwa, a fish and wildlife specialist, and Nancy Morin, a botanist, were asked to serve on the Executive Board with Chairman M. Kosztarab. They accepted this task. I found special pleasure in working for two years with an active committee which included a good representation of experts from different biological disciplines. The planning committee and the advisory board for the national biological survey are now comprised of 21 scientists and administrators representing or serving as resource persons for 15 organizations, 14 disciplines, and 11 government agencies. They are listed at the end of this article.

The scientific community has endorsed the national biological survey concept through 33 organizations representing over 200,000 scientists in the United States (Table 1). During the past two years, members of the planning committee conferred with and obtained a positive response from 11 U.S. government agencies and direct endorsement from three of these. Supporting letters for a national biological survey were also received from 14 leading life scientists and at least five U.S. legislators. I also held meetings with six government agencies, five scientific organizations, and three U.S. legislators; in addition, I gave invitational and/or volunteered oral reports on the national biological survey to 12 scientific/public organizations including the Association of Systematics Collections; Ecological Institute in Lund, Sweden; Entomological Society of America; Hungarian National Museum; 17th International Congress of Entomology (with H. V. Danks); National Research Council, Commission on Life Sciences; Natural Resources Council of America; Staff of the Senate Committee on Environment and Public Works; Sierra Club, New River Group; Virginia Academy of Sciences; and the Entomological Society of Washington.

To disseminate information on the proposed national biological survey during 1984 and 1985, at least 17 news coverages were provided; some through the *Associated Press* by Dorothy Gast (1984, nationwide), three in *Science* (Kosztarab, 1984a and b; Gardner, 1984), and three in *ASC Newsletter* (Edwards, 1984; Kim, 1984; Kosztarab, 1984c). In addition, I gave three radio interviews (Los Angeles) and one television (CBS Roanoke) interview.

Our joint efforts apparently bore fruit with the special interest recently demonstrated in a national biological survey by a number of U.S. legislators, and by the organization of two national meetings on this topic, one held in May of 1985 by the ASC, the other sponsored by the American Institute of Biological Sciences and scheduled for 1986. Since the ASC meeting, the Secretary of the Smithsonian Institution, Robert McC. Adams, has been quoted in the press as being interested in a national biological survey (Holden, 1985).

To remove any doubts as to what a national biological survey could do for this nation, I will summarize its anticipated practical benefits. The survey will:

Table 1. Associations and other organizations endorsing a national biological survey through resolutions or letters.

A. United States or Regional Organizations
1. American Association for the Advancement of Science
2. American Bryological and Lichenological Society
3. American Institute of Biological Sciences
4. American Ornithological Union
5. American Phytopathological Society
6. American Society of Mammalogists
7. American Society of Parasitologists
8. American Society of Plant Taxonomists
9. Association of Southeastern Biologists
10. Association of Systematics Collections
11. Ecological Society of America
12. Entomological Society of America
13. Entomological Society of Washington
14. Lepidopterists' Society
15. Mycological Society of America
16. North American Benthological Society
17. North American Lake Management Society
18. Society for the Study of Amphibians and Reptiles
19. Society of Nematologists
20. Southern California Association of Marine Invertebrate Taxonomists
21. Virginia Academy of Sciences
22. Weed Science Society of America

B. State "Biological" Surveys
1. Biological Survey, New York State Museum
2. Illinois Natural History Survey
3. Kansas Biological Survey
4. North Carolina Biological Survey
5. Ohio Biological Survey
6. Texas System of Natural Laboratories Inc. (A non-profit private organization)

C. International and/or Foreign Organizations
1. Australian Biological Resources Study (Bureau of Flora & Fauna)—Canberra
2. Biological Survey of Canada (Terrestrial Arthropods)—Ottawa
3. Ecological Institute—Lund, Sweden
4. Mexican Academy of Sciences (Academia de la Investigaćion Cientifica)—Mexico City
5. 17th International Congress of Entomology, 1984—Hamburg

1) Provide, for the first time, an inventory of the fauna and flora;

2) Generate needed identification manuals for the components of our biota (the well-illustrated manuals will serve as teaching tools for secondary and higher education);

3) Save money for this nation by providing nationwide coordination and avoiding duplication of related efforts by federal, state, and private agencies;

4) Serve as a catalyst for needed work in this country and focus efforts *first* on the endangered habitats, and *second* on the entire biota;

5) Establish a computerized data bank of our living natural resources and interface it with existing data banks on related subjects (Olson, 1984);

6) Provide periodic reports to the Congress, government agencies, and U.S. corporations on the status of the biota and aid government agencies by supplying essential information on the biota;

7) Promote biological control of pests by supplying adequate taxonomic, biological and distributional information on the natural enemies of pest organisms;

8) Help elucidate the effects of acid rain and other environmental pollutants on the biota while supplying a base for monitoring changes in the composition of the biota;

9) Assist U.S. corporations by providing data on the renewable natural resources of this country, because no such comprehensive data base exists today (Train, 1984);

10) Generate data needed to produce climatic profiles for species, by utilizing records available on the species and on temperature and precipitation of the area (such a system enables forecasting of the distribution of species and is in use in Australia (Bridgewater, 1984));

11) Improve the national security of this country by providing new knowledge on certain natural resources essential for human survival in a national emergency;

12) Enhance international coordination and cooperation on renewable natural resources in North America and elsewhere (Danks and Kosztarab, in press) [a number of recent conferences have called for such international cooperation (Cushwa and Tungstall, 1983)];

13) Help to preserve our genetic and aesthetic natural resources by maintaining biological diversity.

It is apparent from these many anticipated practical applications that a national biological survey is a worthwhile, but extensive undertaking, and the consensus among scientists is that the organization, initiation and implementation of a national biological survey is a broad interdisciplinary, interinstitutional effort that should involve the full range of biological sciences in this country.

The planning committee for the national biological survey project consists of: Ross H. Arnett, Jr., Florida State Collection of Arthropods; *Charles T. Cushwa, Fisheries and Wildlife Science, Virginia Polytechnic Institute and State University (VPI&SU); *Stephen R. Edwards, ASC; *Lafayette F. Frederick, Howard University; *K. C. Kim, Pennsylvania State University; *Nancy Morin, Missouri Botanical Garden; Robert L. Pienkowski, VPI&SU; Paul G. Risser, State Natural History Survey of Illinois; *Amy Y. Rossman, Mycology Laboratory, Biosystematics and Beneficial Insects Institute, USDA; *Thomas E. Wallenmaier, Animal and Plant Health Inspection Service, USDA; and *Michael Kosztarab, VPI&SU, Chairman.

Advisors and/or Agency Consultants are: Peter Bridgewater, Bureau of Flora and Fauna of Australia; Hugh V. Danks, Biological Survey of Canada; Allan Hirsch and Rufus Morison, Environmental Protection Agency; Lloyd Knutson and Howard E. Waterworth, U.S. Department of Agriculture; Robert C. Riley, American Association for the Advancement of Science; **Stanwyn G. Shetler, Smithsonian Institution; Richard N. Smith, Fish and Wildlife Service; and Nevenna Tsanoff Travis, Texas System of Natural Laboratories, Inc. (* indicates present at the initial organizational meeting; ** ex-officio observer with George M. Davis, President of ASC).

I express my heartfelt thanks to the persons listed and to those not listed who in their own way assisted the national biological survey effort.

ACKNOWLEDGEMENTS

The following colleagues reviewed and improved my draft of this article: L. T. Kok of this university, Lloyd Knutson and William L. Murphy at the U. S. Department of Agriculture, and Stanwyn G. Shetler of the Smithsonian Institution.

LITERATURE CITED

Bridgewater, P. 1984. *Australian Biological Resources Study. Annual Report 1983–1984.* Bureau of Flora and Fauna, Dept. Home Affairs and Environment. Canberra. 24 p.

Cameron, J. 1929. *The Bureau of Biological Survey: Its history, activities and organizations.* The John Hopkins Press, Baltimore. 339 p. (Reprinted 1974, Arno Press, New York).

Cushwa, C. T., & D. B. Tungstall. 1983. Wildlife in the United States: The U.S. response to the Organization for Economic Cooperation and Development 1982 wildlife questionnaire. *In: Proc. of Internat. Confer. Renewable Resource Inventories for Monitoring Changes and Trends.* Corvallis, OR, SAF 83–14: 43–44.

Danks, H. V. (ed.) 1978. Canada and its insect fauna. *Mem. Entomol. Soc. Canada* No. 108: 573 p.

Danks, H. V., & M. Kosztarab. In press. Biological surveys. *In: Biosystematic services in entomology.* Proc. of a Symposium held at the 17th International Congress of Entomology, Hamburg. (38 p. ms.)

Edwards, S. R. 1984. A national biological survey. *ASC Newsletter* 12(1): 6.

Gabrielson, I. N. 1940. Bureau of Biological Survey, *In:* Anonymous, *Annual report of the Secretary of the Interior.* U.S. Department of Interior. 528 p.

Gardner, A. L. 1984. Biological survey. *Science* 224: 1384.

Gast, D. 1984. VPI professor urges U.S. biological survey. *Washington Post, Virginia Edition.* July 17, p. B-3.

Holden, C. 1985. New directions for the Smithsonian. *Science* 228: 1512–1513.

Kim, K. C. 1984. Current programs of the Council of Applied Systematics. *ASC Newsletter* 12(4): 29–30.

Kosztarab, M. 1975. Role of systematics collections in pest management. *Bull. Entomol. Soc. Amer.* 21(2): 95–98.

Kosztarab, M. 1984a. A biological survey of the United States. *Science* 223: 443.

Kosztarab, M. 1984b. (Untitled) *Science* 224: 1384. (A response to Gardner's letter).

Kosztarab, M. 1984c. Abstract from the symposium "The systematics community: priorities for the next decade". *ASC Newsletter* 12(6): 55.

Olson, R. J. 1984. *Review of existing environmental and natural resource data bases.* Oak Ridge Nat. Lab., Environmental Sci. Div., Publ. No. 2297: 61 p.

Train, R. E. 1984. Executive summary. *In: Corporate use of information regarding natural resources and environmental quality.* World Wildlife Fund-U.S. 3 p.

Systematics and Long-range Ecologic Research

Barry Chernoff
The Academy of Natural Sciences of Philadelphia

Abstract: A national biological survey is considered in the context of systematic and long-range ecologic research, and some of the fundamental relationships between ecology and systematics are discussed. The degree to which we know the biota of the United States is a function of the included areas because of the discrepancy of knowledge among geographic regions. Knowledge among groups of organisms is also shown to be disproportionate. Moreover, the foundation of knowledge even for well-studied organisms may be rather porous. Some of the benefits of long-term ecological research include the ability to discern pattern from noise and the ability to determine the rather complex, temporally variable, responses of ecosystems. The overwhelming rationale for a national biological survey resides in the necessity to protect our biota. The capacity of our management policies to accomplish this depends ultimately upon sound systematic and ecologic studies.

Keywords: National Biological Survey, Systematics, Ecology, Phylogenetics, Evolution, Conservation, Ecosystem, Cryptic Species.

INTRODUCTION

During the last several decades there has been increasing emphasis by systematists and ecologists on ecosystems outside of the continental United States, especially in tropical areas. Kosztarab (1984) has questioned our knowledge regarding native organisms of the United States, and whether or not a national biological survey is warranted. In this paper, I focus upon the systematic and ecologic imperatives for a national biological survey. My purpose is not to debate the relative importance of acquiring knowledge on faunas, floras and ecosystems within or beyond the borders of the United States. Studies on all such systems are essential, if not critical. For example, as much as 245,000 sq. km of tropical forests is being destroyed annually (Myers, 1979). A national biological survey should not be instituted at the expense of existing programs. Reprogramming will destroy the basis for crucial, world-wide scientific research performed by U.S. scientists; such a result would be both unfortunate and unacceptable.

Perhaps the single most important reason to compel the implementation of a national biological survey is conservation. To ensure the continuance of our native fauna and flora, we need to know the taxa that are present as well as their habits and requirements. Sound management policy must be based upon sound systematic and long-range ecologic studies. At the same time, changing land-use patterns within the United States make the job more difficult as ecosystems become ever more fragmented. A national biological survey would serve as a stimulus and directive for particular scientific endeavors to be carried out locally.

This paper will address four main issues in the context of a national biological survey. 1. What is the fundamental relationship between systematic and ecologic research? 2. How well do we know the organisms of the United States? In general, what is the basis for our systematic knowledge? 3. What are the benefits of long-term ecologic studies? 4. What special conservation considerations must be addressed?

RELATIONSHIPS BETWEEN SYSTEMATICS AND ECOLOGY

Systematics and ecology are two branches of the biological sciences that have their origins in the field of natural history. Systematics and ecology together comprise a large portion of what is today recognized as comparative and evolutionary biology. Both disciplines seek to discover and explain patterns found in the natural world—one relative to genealogies and the other relative to organismic interactions. Alexander (1969) and Eldredge and Cracraft (1980) noted that comparative biology allows us to analyze and capture biotic patterns from which process-oriented theories may be inferred. Systematics and ecology are fundamentally interrelated. Interpretations in each discipline often depend upon information or a frame of reference from the other. In order to understand their relationships we must first understand their definitions.

Systematics has been defined in several ways. During the last two decades, the modern redefinition of the term has been provided by Simpson (1961): "Systematics is the scientific study of the kinds and diversity of organisms and of any and all relationships among them." Mayr (1969) provides a slight shortcut: systematics is the science of the diversity of organisms. Interestingly, neither Simpson (1961) nor Mayr (1969) restricted 'relationship' to a phylogenetic perspective. Rather, they conceive 'relationship' to include, as stated by Mayr (1969), "all biological relationships among organisms." Eldredge and Cracraft (1980) restricted the sense of the term to include only the study of "orderly" or hierarchical patterns found in nature. Wiley (1981) basically agrees with Eldredge and Cracraft, but was more loquacious: "Systematics is the study of organismic diversity as that diversity is relevant to some specified kind of relationship thought to exist among populations, species, or higher taxa." Cracraft and Eldredge, Wiley and others require relationships or hierarchies to be phylogenetic (i.e., to reflect genealogies), and I agree. Nonetheless, I prefer a slightly broader concept of systematics that embodies three aspects: (i) the discovery and identification of organismic diversity; (ii) the estimation of phylogenetic relationships among organisms: and (iii) the study of evolutionary processes that account for such phylogenetic patterns and diversity.

Taxonomy is generally considered to comprise the theory and practice of classifying organisms. Thus, taxonomy is ancillary to systematics, regardless of the particular classification theory (e.g., evolutionary, phenetic or phylogenetic).

There is little, if any, disagreement over the meaning of ecology. The term denotes the interrelationships among organisms and their habitats. Patterns emerge from these interrelationships, whether from interspecific interactions, diversity and complexity of species assemblages, or the flow of energy and nutrients through an ecosystem (Ricklefs, 1973). Patterns then become the objects of study to elucidate the processes underlying them. Exceptions to this general scenario comprise those studies that are purely descriptive and not comparative (e.g., an autecologic study of one species in one locality during one time period): general patterns are not produced and general processes cannot be elucidated.

There are three main relationships between ecology and systematics: (i) process-level inferences depend upon phylogenetic patterns; (ii) phenotypic expression and character covariances are a joint function of genealogical and ecologic factors; and (iii) proper species identifications provide both the basis and predictions for ecologic studies. The first relationship is best viewed in the context of patterns and processes. Process is causally prior to pattern; without process there can be no pattern. For example, ancestor-descendant or parent-offspring relationships give rise to lineages. Inferences about evolutionary and many ecologic processes require patterns as reference systems. Patterns provide directionality for studies of processes—a change in the perceived pattern usually requires a change in the conclusion about process (Lauder, 1981; Fink, 1982). Consider the heterochronic processes of development that comprise paedomorphosis. Without an estimate of phylogeny, there is no way a priori to distinguish paedomorphosis from a mere primitive condition. That is, to infer that developmental sequences were somehow truncated or deleted, one needs to establish that the sequences were already present in the lineage (for further discussion see Fink, 1982, and Bookstein et al., 1985, in comparison to Alberch et al., 1979).

Gould and Vrba (1982) have discussed the dependence of adaptation as a conclusion upon the phylogenetic distribution of some trait. The morphological, physiological or behavioral traits that allow a species to survive and reproduce successfully in particular environments are *not* adaptations of the species if the traits were present in the common ancestor of the species and its closest relative. That is, organisms may be able to meet the demands of their environments by coopting ancestral adaptations. For example, Smith (1981) has argued that desert pupfishes (genus *Cyprinodon*) are not necessarily adapted to desert conditions; rather, they survive because of ancestral physiological tolerances to harsh estuarine environments.

Treatments of other topics in the ecologic or evolutionary literature have also been weakened by lack of attention to phylogenetic patterns. For example, genealogical information is crucial to inferences about taxon cycles of the West Indian avifauna (Ricklefs and Cox, 1972); their model requires that the end products of the cycle are in fact the most derived members of their lineages. Arguments for or against ecologic character displacement in Darwin's finches should include a genealogic perspective, but they do not (Grant and Schluter, 1984; Simberloff, 1984; and references therein; although Simberloff, 1983 came close to admitting this point). Lastly, Brooks (1979), Brooks et al. (1981), and Mitter and Brooks (1983) showed that inferences about host-parasite coevolution are determined most appropriately from the evolutionary relationships of both hosts and their parasites.

The second relationship acknowledges the joint effects of genealogic or ecologic factors upon character expression and covariance. The phenotypes of a species and their covariances may be partitioned into historic (i.e., genealogic, phylogenetic) and non-historic (i.e., ecologic, environmental) components, plus random noise (uncorrelated variation and measurement error). In order to account for the ecologic component of character expression or covariance we must first factor out the historic component; the converse is also true. Non-genealogical, environmental or ecologic affects upon character expression are well documented across a broad spectrum of organisms (e.g., Gilbert, 1966; Kreuger and Dodson, 1981; Chernoff, 1982; James, 1983; Rathke, 1984; and references therein). These processes may have such an overriding effect upon character expression that the phylogenetic signal is obscured (Chernoff, 1982). Conversely, phylogenetic constraints upon ecologic expression must also be accounted for before the proper ecologic signal can be analyzed (e.g., Dunham et al., 1979; Stearns, 1984, but see Dunham and Miles, 1985, for corrections). The ecology of an organism is, in part, a function of its phenotype, as its phenotype is, in part, a function of its ecology. Neither systematists nor ecologists should lose sight of this fundamental dependence.

The third and most obvious relationship between ecology and systematics concerns species recognition and identification. Ecologic data, such as preferences of food, habitats, breeding sites and times, etc., can provide important insights into the composition and population structure of species. Precise delineation of species is particularly important for ecologic or environmental research. For example, a polychaete worm of the genus *Capitella* had long been used as an ecologic standard for marine pollution studies until Grassle and Grassle (1976) discovered allozymically that what was thought to be one species was actually six. Each of the six had differing life histories, thus confounding former studies on the natural variation of a single species.

The last two points raise the issue of reciprocal illumination or predictivity. To a certain extent systematic or ecologic findings can provide predictions for the other discipline. For example, Eberhard (1982) used the characteristics of the webs and web-building behaviors to make predictions about the relationships of orb-weaving spiders. Similarly, variation in life-history traits, in part, has led Hoagland (1984) to question the integrity of the gastropod species, *Crepidula* "*convexa*". Systematists' interpretations of species boundaries are generally used by ecologists as the basis for distinguishing their units of study. Elevation to specific status of once-synonymized species may serve as a prediction of ecologic difference (e.g., habitat and feeding preferences and interspecific competition were discovered between two species of silverside fishes by Lucas, 1978, after the second species was recognized as distinct from the first by Johnson, 1975). An hypothesis related to the ecology of some species could also be formulated from the phylogenetic distribution of the ecologies of its relatives. For example, an organism that is most closely related to a group of sand burrowers might also be expected to be a sand burrower.

There are, however, limitations to the extent to which predictions across disciplines can be made. Pitfalls can result from variations or polymorphisms of ecologic characteristics that are not indicative of species-level differences. For example, environmental conditions can lead to semelparous or iteroparous reproductive cycles among populations of a species (Leggett and Carscadden, 1978), to cyclical parthenogenesis (Lynch, 1984), or to alternative mating strategies (Dominey, 1984). Nonetheless, ecologic variations or polymorphisms exhibited by a species can be discerned with the same attention given to variation in morphologic or genetic traits.

A more insidious pitfall inheres in the assumption that members of higher taxa (e.g., genera, families) share common ecologies or inhabit the same ecologic niche or 'adaptive zone'. This assumption is founded in the belief that families and genera, or higher taxa in general, "owe their origin to the invasion of this zone by a founder species and to the subsequent active and *adaptive radiation* which usually follows a successful adaptive shift..." (Mayr, 1969; emphasis mine). Mayr's (1969) definition of adaptive radiation ("Evolutionary divergence of members of a single phyletic line into a series of rather different niches or adaptive zones.") thus, contradicts his notions of ecologic similarity within a higher taxon. But even beyond this difficulty with definitions, there seems to be little evidence that members of higher taxa have common ecologies (Selander, 1969). For example, consider the different feeding and nesting ecologies of the Galapagos finches within each of the genera *Geospiza* or *Camarhyncus* (Lack, 1947). The sandpiper genus *Calidris* contains species that feed in the uplands and others that feed in shallow open water (J. P. Myers, pers. comm.). The atherinid fish genus *Atherinella*, a monophyletic group, contains species that strain plankton, others that are strictly piscivorous, as well as purely marine species and species that live in high elevation streams (Chernoff, 1983). In the euphorbaceous plant genus *Macaranga*, many species are strictly early-succession light-demanding pioneers, while others are shade-requiring late-climax rainforest species (Whitmore, 1973). Many other examples could be cited. Unfortunately, this ecologic-similarity criterion has been applied to the formation of higher categories with the result of overly-split or unnatural groups. For example, the crab-plover (*Dromas*), and the pratincoles and coursers (*Glareola, Stiltia, Rhinoptilus*) have usually been allied with the plovers, in part because of common ecologies; DNA-DNA hybridization data show instead that the crab-plover, pratincoles and coursers are most closely related to gulls, terns, jaegers, skimmers and auks (Sibley and Ahlquist, in press). The use of ecologic characters in the formation of systematic relationships should, at best, be viewed with the same suspicion as morphological characters (Cain, 1959; Selander, 1969). Members of higher taxa need only be monophyletic (i.e., each other's closest relatives). The ability to predict the ecology of a taxon should thus be based upon the phylogenetic distribution of ecologies, rather than on an arbitrary assumption of similarity. This point has important consequences for managing ecosystems and will be discussed in the conservation section below.

SYSTEMATIC IMPERATIVE

> "Scientific truth is not a dogma, good for eternity, but a temporal quantitative entity...the time spans of scientific truths are an inverse function of the intensity of scientific effort." Robert M. Pirsig (1974)

The systematic imperative for a national biological survey can be supported by two lines of reasoning. The first derives from how well the fauna and flora of the United States are known. I will elaborate this aspect from the following perspectives: (i) the geographic limits of the national biological survey; (ii) the distribution of scientific effort across different groups of organisms; and (iii) the relation between cryptic species and modern systematic techniques. The second reason concerns the rate of man-induced extinctions. This latter topic will be addressed in the section on conservation.

Our knowledge of the biota of the United States is in part dependent upon the geography that we choose to consider. In many ways we know the organisms in the northeastern U.S. better than we do those of the Southwest, and far better than those that inhabit Hawaii. A national biological survey should be just that: national. It should, therefore, include the fauna and flora of the continental United States, Alaska, Hawaii, and the oceans and bottoms contained within U.S. territorial waters. Whether or not the survey should include U.S. territories and protectorates (e.g., American Samoa, Guam, Puerto Rico, and the Virgin Islands) is another matter and beyond the scope of this paper.

To demonstrate better the relationship between knowledge and geography, consider the Orthoptera of the U.S. (D. Otte, pers. comm.). If we accept current schemes, possibly less than 20% of the continental U.S. fauna remains undescribed. For Hawaiian crickets, however, there were 30 described species as of 1950; now there are about 200 described forms and the number could rise to as many as 1,000 species. In general, the inclusion of Hawaii will lower the level of our knowledge regarding many groups of organisms. The Hawaiian example serves to illustrate why provisions should be made to target relatively unknown regions for special study. Intensive efforts in such regions would provide a large bulk of critical information in an efficient manner.

There are two ways of viewing the current state of knowledge regarding the biota of the United States, independent of geography. One involves estimates of what is described versus what we estimate to exist; the other is to focus upon the relative information available among groups of organisms. Estimating the size of the biota both for particular systematic groups (e.g., chironomids) or for the United States as a whole is almost an impossible task. To do so requires extrapolations for which we have no data. How can we estimate the numbers of organisms that we have yet to discover? This is different from known taxa which are merely *undescribed.* Such species have been discovered, identified and only lack formal description. For example, there is a list of undescribed fishes from the United States (Jenkins, 1976), many of which have been known to ichthyologists for decades. Estimates of the size of the U.S. ichthyofauna would include these undescribed but known species but would not include the new taxa we may someday discover.

What we do know about the U.S. biota, even on a cursory level, is incommensurate across groups of organisms. As a result there are huge gaps in the published systematic coverage of the biota. For some groups, such as birds, there have been intensive studies on the species of the United States and our knowledge appears to be reasonably complete. For other groups, much exploration is still required. For example, Hodges (1976) noted that even for certain groups of insects, such as Symphyta or Formicidae, where more than 75% of adult specimens from North America can be assigned to described species, less than 10% of the larval specimens can be assigned. Between 400 and 600 new diatom taxa have been described each year since 1965 (C. W. Reimer, pers. comm.), and there is little indication that the trend will diminish. If anything, the application of scanning electron microscopy (e.g., Qi et al., 1982) and multivariate statistics (e.g., Theriot and Stoermer, 1982) will likely increase the diatom flora. Morphologic and genetic studies of daphnids demonstrate clearly that several recognized species are polyspecific (e.g., Hebert, 1978; Hebert and Crease, 1980). When such "well studied" organisms as *Daphnia magna* or *D. pulex* present fundamental taxonomic problems, the systematic imperative for a national biological survey becomes ever more clear. Nonetheless, there are groups for which we lack even the fundamental taxonomic knowledge to know where the problems lie. Major systematic efforts are needed in groups such as rotifers, nematodes, diptera, hemipterans, beetles, mites, mistletoes and many, many more. Thus we should ensure, by way of funding, that systematic studies of the poorly known groups will be undertaken.

Beyond the poorly known groups, our systematic knowledge for many other taxa is too inadequate or in such a state of turmoil that published taxonomies are wholly unacceptable. An important case in point is that of the relatively common freshwater unionid bivalves of North America (ca. 300 species in the U.S.; A. Bogan, pers. comm.). Starnes and Bogan (1982) have noted that for the fauna inhabiting a tributary of the Cumberland River, Kentucky, approximately 70% have undergone taxonomic revision since 1914. Researchers are only now beginning to understand previous taxonomies, compare types (where extant) and apply names on a consistent scientific basis (Starnes and Bogan, 1982; Bogan and Parmalee, 1983; Bogan et al., 1984). The taxonomic chaos for unionids has arisen, in part, because of inconsistent acceptance or rejection of Rafinesque's nomenclature (he described 34 genera and subgenera, 124 species and 55 varieties; Bogan et al., 1984). Much work remains to be done before a well formulated taxonomy can be applied. Hodges (1976) has described serious limitations to the nomenclature applied to species of Lepidoptera. Apparently Edward Meyrick, who died in 1938, described more than 15,000 species but "refused to use a microscope until his later years, and he refused to acknowledge that genital characters where worthwhile. He based much of his classification on the venation as seen through a hand lens." The Compositae (Asteraceae) is one of the three or four largest families of plants with 12,000 to 14,000 species worldwide. The Compositae contains perhaps 1,500 to 2,000 species within the U.S. (B. Stone, pers. comm.), and is exceedingly abundant in the Western U.S.; these are the sunflowers, asters, daisies, etc. Despite their ubiquity and familiarity, the classificaion of the Compositae is being challenged seriously, especially at the generic level (see symposium in *Taxon*, vol. 34, no.1, February, 1985; note papers by Cronquist, 1985, and Funk, 1985). Important realignments and taxonomic change within the Compositae are being proposed; use of the present classification will only be misleading. Recent studies utilizing soft anatomy, multivariate analysis and molecular genetics are providing significant changes in molluscan systematics (e.g., Davis et al., 1982; Davis, 1983). The last example to demonstrate that not all common organisms have stable or ecologically-usable taxonomies is that of the leopard frogs of North America. As noted by Hillis et al. (1983), the *Rana pipiens* complex has been a modern biological enigma. Before 1900, 12 species were described; in the 1940's and 1950's only four species were recognized; today the complex contains more than 20 species, eight of which are in the process of being described (Hillis et al., 1983). Nonetheless, new species of ranids are still being discovered within the United States (e.g., *Rana okaloosae* Moler, 1985). Problems like these may be solved, in part, by more synthetic or monographic treatments. Even regional taxonomies and keys would provide aid to those working with the biota. As Slater (1981) lamented: "After 100 years of distinguished work in Entomology one really still does not know the distribution within a state of most of the insect fauna. There is still no 'Insects of Iowa' series in being or in progress to my knowledge."

The inadequacy of our systematic knowledge may be more pervasive than indicated by the few examples given above. Evaluations of species boundaries should attend to variability: of individuals within populations, among populations and of the characters overall. In certain instances character heritabilities may need to be examined in order to determine the meaning of the variation (Chernoff, 1982). Because the philosophy of systematics is always changing, we have inherited taxonomies for many groups largely devoid of attention to variability or with narrower concepts of species limits. For example in 1859, Charles Girard described the channel catfish, *Ictalurus punctatus*, four times; three of the descriptions appear in the same paper. Furthermore, there are all too many examples, older and modern, of systematic decisions focused upon one or several "key" characters. As Hodges (1976) observed for the systematics of Lepidoptera: "To my knowledge species vary in nearly all characters, and for this reason the male or female genitalia sometimes are no more final for specific determination than the shading of the color pattern, wing length, or other characters." The point is that because of the enormity of the task facing systematists, adequate assessment of variability for all taxa will never be realized. As such, geographic variants, biotypes, polymorphisms and environmentally induced characteristics will result in an over-estimation of the number of valid species.

The last aspects of the systematic imperative for a national biological survey involve cryptic species and new technologies. As Pirsig's insight suggests, some of the "species" we accept at face value may dissolve under closer scrutiny. Thus we must deal with cryptic species, which have little to do with mimicry or crypsis. Rather, cryptic species have been defined as those which are difficult to recognize on the basis of generally used morphological criteria (Walker, 1964). Walker goes on to note that: "The term *sibling species* is frequently used for the same phenomenon, but '*sibling*' connotes a more recent common ancestry than is the case of species that are not '*sibling*.' Since morphological indistinctness and recentness of common ancestry are not perfectly correlated, '*cryptic*' is more descriptive than '*sibling*'..." [italics his]. New technologies (i.e., scanning electron microscopes, statistical analyses, molecular genetics) allow us to scrutinize our organisms ever more closely and often lead to the discovery of cryptic species. Cryptic species are also discovered, however, when systematists go beyond the usual criteria in examining their organisms, whether the criteria are morphological, biochemical, behavioral, etc. Davis (1983) concluded that convergent evolution of the molluscan shell has masked the genetic and anatomical differences among many species.

It is impossible to estimate how many of the species presently recognized are actually composites of two or more cryptic species. However, Walker (1964) states that nearly one-fourth of the species of ensiferan Orthoptera are cryptic. Whether this high percentage is applicable to other groups remains to be demonstrated. Recognition of cryptic species seems to be independent of a systematist's definition of the species concept because the evidence obtained is usually so persuasive that little doubt remains. An interesting generality emerges from the various examples of cryptic species: once the "hidden" species-defining characteristics have been discovered, subtle but consistent differences among "conventional" traits frequently become recognizable. Cryptic species or forms can be found among all groups of organisms. Examples range from the more obvious to the subtle, and have been discovered using such characteristics as genitalic morphology (Burns, 1984), calling songs (Walker, 1964), steam-volatile terpenoids (Adams, 1983), osteology and shape analysis (Chernoff et al., 1982), allozymes (Davis and Fuller, 1981; Davis, 1983; Hoagland, 1984), 2-D protein patterns and isozymes (Ferris et al., 1985; Huettel et al., 1984), and early life stages (Hoagland, 1984).

In conclusion, a national biological survey is needed from a systematic perspective because fundamental information is lacking. The application of effort on the biota of the United States is misapplied, and basic knowledge across groups of organisms is very inconsistent. As a result there are groups of organisms and geographic regions for which our knowledge is poor to non-existent. Work is needed in some better-studied taxa because the existing systematics are unreliable overall. Lastly, general systematic surveys are needed that employ methods (e.g., molecular genetics, morphometrics) sensitive enough to identify cryptic species. At the very least, general collections should be archived to allow for traditional and new approaches. Studies not now conceived may be important in the future.

LONG-RANGE ECOLOGIC RESEARCH

The ecologic necessities for a national biological survey are as compelling as those from a systematic perspective. Some ecologic research can be performed in the laboratory, but most studies depend upon active programs carried out in the field. As the number of undisturbed field sites in the United States diminishes, field programs become ever more hampered. Yet our knowledge of the ecologies and life histories of the biota of the United States is far from complete. There are even agricultural pests for which adequate life history information does not exist.

Like systematics, ecology is a discipline of hierarchic levels. For example, studies can be directed at the level of the single species, species interactions, community organization, behavior and evolution of entire ecosystems, or ecosystem interactions. It is not sufficient to know the names, relationships and distribution of organisms, we must also know their biologies: that is, how they interact with their physical environment, what their population biologies are, and how they interact with other species (e.g., competition, predation, food source). Although I am not an ecologist, I suspect that like systematics, our ecologic knowledge across groups of organisms, geographic regions and ecosystems is inconsistent.

The question at hand, though, concerns the necessity for long-range ecologic research. The length of time devoted to long range studies should be scaled to the needs of the particular question in relation to the response time of the organisms or the ecosystem (Tilman, 1982)—some systems may require only hours (e.g., bacteria) while others may require years (e.g., trees). Two of the most fundamental considerations, however, are that the studies be of sufficient duration to: (i) decipher pattern from noise, and (ii) distinguish the total shape of the pattern (Likens et al., 1977; Bormann and Likens, 1979).

Long-term research on the Hubbard Brook ecosystem in the northeastern United States (Likens, et al., 1977; Bormann and Likens, 1979) provides beautiful examples of how ecologic conclusions would change if sampled over a shorter time-span. Consider their compilation of the responses of a northern hardwood forest to clear-cutting (Figure 1). If one were to conduct short term studies (e.g., one to three years), one would arrive at very different conclusions dependent upon how many years after clear-cutting the studies were performed. Biological systems are complex and do not necessarily respond linearly with respect to time (e.g., see Mooney and Gulmon, 1983; and Gill, 1980). Often biological systems present an initial lagged response. For example, Gaylord and Hansen (1983) found the response-time of trout productivity to lag three to four years after the sediment load of streams had been increased; had they quit after two years they would have concluded erroneously that increasing sediment load had no significant effect.

The Long-Term Ecological Research Program was implemented by the National Science Foundation to provide a solution to the problems mentioned above as well as to alleviate the counterproductive activities of competing for funding on a short-term basis. The rationale, background, implementation and goals of the Long-Term Ecological Research Program is given in a cogent article by Callahan (1984). The program now supports research at 11 sites comprising the following habitats: coniferous forest, oak savannah, deciduous forest, salt marsh—estuary, desert, tall-grass prairie, large rivers, alpine tundra—lakes, northern lakes, fresh-water swamp, and short-grass prairie. Most notable among ecosystems not represented for long-term research are marine communities, especially coral reefs. Callahan (1984) noted that the successful proposals, to date, could be categorized as follows: i) "the effects of physical environmental variables on the structure and the change in the structure of biotic communities;" ii) "the processes by which herbivorous populations are regulated;" iii) "the processes that regulate the rates of accumulation and transport of decomposing organic matter;" iv) "the processes that influence the rates at which inorganic nutrients are taken up, utilized, and released by the biota;" and v) "the role played by major disturbances in maintaining or changing the character of ecosystems."

The examples and arguments presented above demonstrate clearly that long-range ecologic research should continue to be implemented and expanded on the various ecosystems and organisms in the United States. Shorter-term studies, while valuable as pilots, can lead to erroneous conclusions when extrapolated to a longer time span. Such errors could have grave consequences for management and conservation policies.

CONSERVATION AND CONCLUSIONS

> "One of the most profound developments in the application of ecology to biological conservation has been the recognition that virtually all natural habitats or reserves are destined to resemble islands...." Wilcox (1980)

This paper has presented the systematic and ecologic imperatives for a national

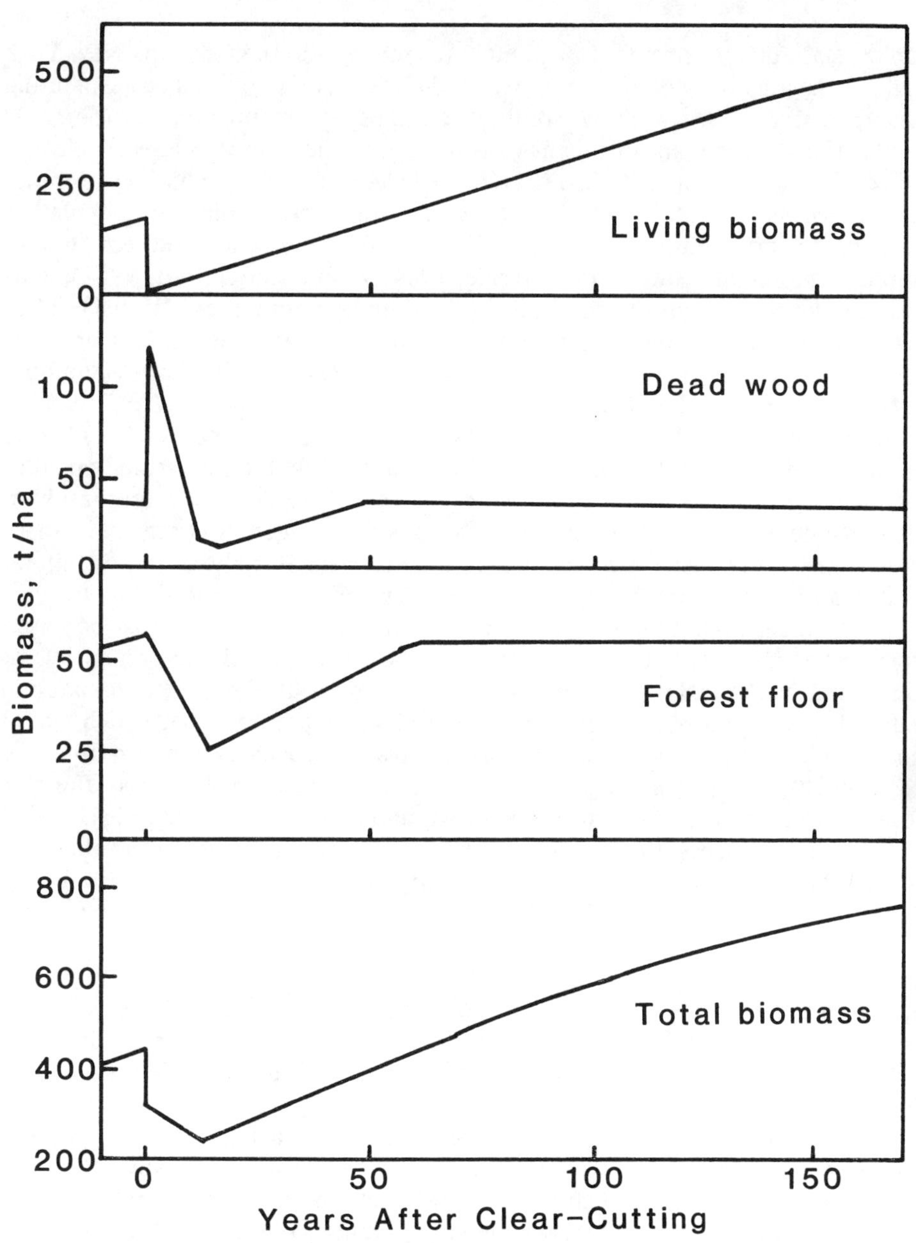

Fig. 1. Schematic of response of northern hardwood forest to clear-cutting; modified from Bormann and Likens (1979: figure 1–10).

biological survey. Recommendations based upon the arguments given above are presented in the next section. I would like to present briefly some of the conservation implications of a survey, because conservation may be the most compelling reason for the institution of such a survey (see also Kosztarab, 1984).

As growth and development continually fragment U.S. wilderness areas, the theories of island biogeography and extinction take on a sudden urgency for the biota. The thought of natural habitats as islands even applies to ocean bottoms and coastal waters. Consider for example the sewage dumped into the New York Bight in the migratory path of flounders and lobsters, or a polluted bay separating cleaner waters to the north and south. The ecology of fragmented environments is precarious and not all organisms can adapt to limited home ranges (Soule and Wilcox, 1980; Whitcomb et al., 1981). Knowledge of population biology becomes critical because survival depends upon such parameters as effective population size, growth and maturation rates, density-dependent mortality, and recruitment. Recent experimental studies (e.g., Abele, 1984; Wilbur and Travis, 1984) demonstrate the changes and instabilities in community composition of "island" or "fragmented" areas. Research is needed to discover what minimum areas or combination of fragments are necessary to conserve natural diversity (Diamond, 1980; Whitcomb et al., 1981).

At the same time, changing habitats and species compositions demonstrate the immediate need for systematic studies. Man-induced ecologic pressures can alter the phenotypes and genotypes of organisms (e.g., *Biston betularia*; see Franklin, 1980). Systematists will have to seriously consider the consequences of changes in lineages of organisms that have occurred since habitat alteration. Temporal studies and comparisons with older museum specimens are crucial.

Management policies designed to conserve the national biota must be based upon thorough systematic and ecologic studies. Counter to the contention of Hirsch (this volume), I believe it is impossible to "overstate the need for taxonomic data." In an absolute sense, how can we manage or preserve that which we do not know? The selective component or exemplar approach to conservation may be affordable, but is nonetheless flawed. Decisions made on the basis of one or several components (organisms) of an ecosystem tend to minimize or exclude all the other components. Reconsider the above discussion concerning the ability to predict the ecologies of related species based upon generic- or family-level classification. If one of the exemplars used for managing a coastal shore zone includes only open-water feeding species of sandpiper, then subsequent decisions may endanger the upland feeding species. The vulnerability of the target-organism approach also depends entirely upon substantial systematic data. Reconsider the example of the marine polychaete worm that was used as a primary indicator of pollution—it turned out to be a complex of six species (Grassle and Grassle, 1976). Former predictions based upon variation in life history were thus rendered baseless. The ability to identify all the organisms within a region, to know their ecologies and interactions, will probably never be achieved. However, we still must admit that decisions based upon data, as complete as possible, will be sounder than the selective target-organism approach. What available funds will allow us to accomplish differs significantly from what is truly "best". This distinction should always be acknowledged.

The consequences of the changing landscape in the United States, and the rest of the world, are embodied in two words: endangerment and extinction. There are too few regional studies on endangered and extinct species (e.g., Linzey, 1979). The current species survival status of certain groups such as large mammals, birds and fishes have been closely monitored by specialists (e.g., Myers, 1979; or Ono et al., 1983, among others); their efforts must continue. But what of the more obscure groups of organisms? How fast are they disappearing? A national biological survey would contribute importantly to answering these questions and thus help in conserving our biota.

RECOMMENDATIONS

1. A national biological survey should include the continental United States, Alaska, Hawaii and oceans and bottoms within U.S. territorial limits. Geopolitical limitations should not be imposed on studies of organisms that range naturally beyond U.S. borders.
2. Efforts should be taken to ensure systematic and ecologic coverage of poorly known groups and poorly studied regions.
3. Monographic or synthetic studies should be encouraged, as well as the production of regional faunas and floras, including keys. Systematic studies should, when possible, combine comprehensive morphological, biochemical and molecular genetic analyses.
4. A national biological survey should incorporate a comprehensive archival program that will accommodate morphological and biochemical analyses, and to the extent practical, also include photographic records. This program should archive all life-history stages of organisms, with particular attention to the early stages.
5. A national biological survey should integrate systematic and ecologic information by funding multidisciplinary teams.
6. Additional ecosystems should be selected for enhanced long-term ecologic and systematic research. The areas should be large enough to incorporate experimental and control subregions. The research activity should be intensive, with multidisciplinary teams. The ecologic effects of ecosystem fragmentation should be considered.

ACKNOWLEDGEMENTS

Many people contributed in important ways to the completion of this manuscript; I am most grateful to all of them. The following provided literature and took time from their valuable research to discuss aspects of this paper with me: A. Bogan, G. M. Davis, F. B. Gill, C. E. Goulden, J. A. Hendrickson, K. E. Hoagland, L. Knutson, R. E. Moeller, J. P. Myers, D. Otte, C. W. Reimer, C. L. Smart, W. F. Smith-Vaniz, B. Stone, A. Tessier, and T. Uzzell. L. Knutson, W. F. Smith-Vaniz, B. Stone, and T. Uzzell provided many important criticisms of the manuscript. I would like to thank the conference organizers and editors of the volume, K. C. Kim and L. Knutson, as well as the Association of Systematics Collections for hosting the symposium and providing the vehicle for its publication. I am also grateful to President T. P. Bennett, M. B. Fischer, and G. M. Davis of the Academy of Natural Sciences of Philadelphia for helping to make my participation possible.

LITERATURE CITED

Abele, L. G. 1984. Biogeography, colonization, and experimental community structure of coral-associated crustaceans. *In*: D. R. Strong et al. (eds.) *Ecological communities: conceptual issues and the evidence.* Princeton Univ. Press, New Jersey.

Adams, R. P. 1983. Infraspecific terpenoid variation in *Juniperus scopulorum*: evidence for Pleistocene refugia and recolonization in western North America. *Taxon* 32: 30–46.

Alberch, P., S. J. Gould, G. F. Oster & D. B. Wake. 1979. Size and shape in ontogeny and phylogeny. *Paleobiology* 5: 296–317.

Alexander, G. R. & E. A. Hansen. 1983. Effects of sand bed load sediment on a brooktrout population. *Michigan Dept. Nat. Resources Fish. Research Report No. 1906*, 50 p.

Alexander, R. D. 1969. Comparative animal behavior and systematics. *In: Systematic biology. Proceedings of an international conference.* Nat. Acad. of Sci. Publ. 1692, Washington, DC.

Bogan, A. E. & P. W. Parmalee. 1983. *The Mollusks. Tennessee's Rare Wildlife, Volume II.* Tenn. Wildl. Resources Agency. 123 p.

Bogan, A. E., L. B. Starnes & J. D. Williams. 1984. An examination of some C. S. Rafinesque North American unionid taxa (Bivalvia: Unionidae). *Am. Malacol. Union Inc. Bull.* 3: 105–106.

Bookstein, F. L., B. Chernoff, R. L. Elder, J. M. Humphries, Jr., G. R. Smith & R. E. Strauss. 1985. *Morphometrics in evolutionary biology. The geometry of size and shape change, with examples from fishes.* Acad. Nat. Sci., Philadelphia, Special Publ. No. 15. 277 p.

Bormann, F. H. & G. E. Likens. 1979. *Pattern and process in a forested ecosystem.* Springer-Verlag, New York. 253 p.

Brooks, D. R. 1979. Testing the context and extent of host-parasite coevolution. *Syst. Zool.* 28: 299–307.

Brooks, D. R., T. B. Thorson & M. A. Mayes. 1981. Freshwater stingrays (Potamotrygonidae) and their helminth parasites: testing hypotheses of evolution and coevolution. *Adv. in Cladistics, Proc. First Meet. Willi Hennig Soc.* 1: 147–175.

Burns, J. M. 1984. Evolutionary differentiation: differentiating gold-banded skippers—*Autochton cellus* and more (Lepidoptera: Hesperiidae: Pyrginae). *Smithson. Contrib. Zool. No. 405.* 38 p.

Cain, A. J. 1959. Taxonomic concepts. *Ibis* 101: 302–318.

Callahan, J. T. 1984. Long-term ecological research. *Bioscience* 34: 363–367.

Chernoff, B. 1982. Character variation among populations and the analysis of biogeography. *Am. Zool.* 22: 425–439.

Chernoff, B. 1983. *Revision of American atherinid fishes, subfamily Menidiinae, and systematics of the subgenus* Atherinella. Unpubl. Ph.D. Dissert., Univ. Michigan, Ann Arbor. 569 p.

Chernoff, B., R. R. Miller & C. R. Gilbert. 1982. *Notropis orca* and *Notropis simus*, cyprinid fishes from the American southwest, with description of a new subspecies. *Occas. Pap. Mus. Zool. Univ. Mich.* No. 698. 49 p.

Cronquist, A. 1985. History of generic concepts in the Compositae. *Taxon* 34: 6–10.

Davis, G. M. 1983. Relative roles of molecular genetics, anatomy, morphometrics and ecology in assessing relationships among North American Unionidae (Bivalvia). *In*: G. S. Oxford & D. Rollinson (eds.), *Protein polymorphism: adaptive and taxonomic significance.* Systematics Association Special Vol. No. 24, Academic Press, New York.

Davis, G. M. & S. L. H. Fuller. 1981. Genetic relationships among recent Unionaceae (Bivalvia) of North America. *Malacologia* 20: 217–253.

Davis, G. M., M. Mazurkiewicz & M. Mandracchia. 1982. *Spurwinkia*: Morphology, systematics, and ecology of a new genus of North American marshland Hydrobiidae (Mollusca: Gastropoda). *Proc. Acad. Nat. Sci. Phila.* 134: 143–177.

Diamond, J. M. 1980. Patchy distributions of tropical birds. *In*: M. E. Soule & B. A. Wilcox (eds.), *Conservation biology.* Sinauer Assoc., Massachusetts. 395 p.

Dominey, W. J. 1984. Alternative mating strategies and evolutionarily stable strategies. *Amer. Zool.* 24: 385–396.

Dunham, A. E. & D. B. Miles. 1985. Patterns of covariation in life history traits of squamate reptiles: the effects of size and phylogeny reconsidered. *Am. Nat.* 126: in press.

Dunham, A. E., G. R. Smith & J. N. Taylor. 1979. Evidence for ecological character displacement in western American catostomid fishes. *Evolution* 33: 877–896.

Eberhard, W. G. 1982. Behavioral characters for the higher classification of orb-weaving spiders. *Evolution* 36: 1067–1095.

Eldredge, N. & J. Cracraft. 1980. *Phylogenetic patterns and the evolutionary process.* Columbia Univ. Press, New York. 349 p.

Ferris, V. R., J. M. Ferris & L. L. Murdock. 1985. Two-dimensional protein patterns in *Heterodera glycines. J. Nematol.* 17: in press.

Fink, W. L. 1982. The conceptual relationship between ontogeny and phylogeny. *Paleobiology* 8: 254–264.

Franklin, I. R. 1980. Evolutionary change in small populations. *In*: M. E. Soule & B. A. Wilcox (eds.), *Conservation biology*. Sinauer Assoc., Massachusetts. 395 p.

Funk, V. A. 1985. Cladistics and generic concepts in the Compositae. *Taxon* 34: 72–80.

Gilbert, J. J. 1966. Rotifer ecology and embryological induction. *Science* 151(3715): 1234–1237.

Gill, F. B. 1980. Historical aspects of hybridization between blue-winged and golden-winged warblers. *Auk* 97: 1–18.

Gould, S. J. & E. Vrba. 1982. Exaptation, a missing term in the science of form. *Paleobiology* 8: 4–15.

Grant, P. & D. Schluter. 1984. Interspecific competition inferred from patterns of guild structure. *In*: D. R. Strong et al., (eds.) *Ecological communities: conceptual issues and the evidence*. Princeton Univ. Press, New Jersey.

Grassle, J. P. & J. F. Grassle. 1976. Sibling species in the marine pollution indicator *Capitella* (Polychaeta). *Science* 192: 567–569.

Hebert, P. D. N. 1978. The population biology of *Daphnia* (Crustacea, Daphnidae). *Biol. Rev.* 53: 387–426.

Hebert, P. D. N. & T. J. Crease. 1980. Clonal coexistence in *Daphnia pulex* (Leydig): another planktonic paradox. *Science* 207: 1363–1365.

Hillis, D. M., J. S. Frost & D. A. Wright. 1983. Phylogeny and biogeography of the *Rana pipiens* complex: a biochemical evaluation. *Syst. Zool.* 32: 132–143.

Hoagland, K. E. 1984. Use of molecular genetics to distinguish species of the gastropod genus *Crepidula* (Prosobranchia: Calyptraeidae). *Malacologia* 25: 607–628.

Hodges, R. W. 1976. Presidential address 1976—What insects can we identify? *J. Lepid. Soc.* 30: 245–251.

Huettel, R. N., D. W. Dickson & D. T. Kaplan. 1984. *Radopholus citrophilus* sp. n. (Nematoda), a sibling species of *Radopholus similis*. *Proc. Helminthol. Soc. Wash.* 51: 32–35.

James, F. C. 1983. Environmental component of morphological differentiation in birds. *Science* 221: 184–186.

Jenkins, R. E. 1976. A list of undescribed freshwater fish species of continental United States and Canada, with additions to the 1970 checklist. *Copeia* 1976: 642–644.

Johnson, M. S. 1975. Biochemical systematics of the atherinid genus *Menidia*. *Copeia* 1975: 662–691.

Kosztarab, M. 1984. A biological survey of the United States. *Science* 223(4635): 443.

Krueger, D. A. & S. I. Dodson. 1981. Embryological induction and predation ecology in *Daphnia pulex*. *Limnol. Oceanogr.* 26: 219–223.

Lack, D. 1947. *Darwin's finches*. Cambridge Univ. Press, London. 208 p.

Lauder, G. V. 1981. Form and function: structural analysis in evolutionary morphology. *Paleobiology* 7: 430–442.

Leggett, W. C. & J. E. Carscadden. 1978. Latitudinal variation in reproductive characteristics of American shad (*Alosa sapidissima*): evidence for population specific life history strategies in fish. *J. Fish. Res. Board Can.* 35: 1469–1478.

Likens, G. E., F. H. Bormann, R. S. Pierce, J. S. Eaton & N. M. Johnson. 1977. *Biogeochemistry of a forested ecosystem*. Springer-Verlag, New York. 146 p.

Linzey, D. W. (ed.) 1979. *Endangered and threatened plants and animals of Virginia*. Virginia Polytechnic and State Univ., Blacksburg. 665 p.

Lucas, J. R. 1978. *Feeding and competition between the silversides* Menidia beryllina (*Cope*) *and* Menidia peninsulae (*Goode & Bean*) *at Crystal River, Florida*. Unpubl. M.S. Thesis, Univ. Florida, Gainesville.

Lynch, M. 1984. The genetic structure of a cyclical parthenogen. *Evolution* 38: 186–203.

Mayr, E. 1969. *Principles of systematic zoology*. McGraw-Hill, Inc., New York. 428 p.

Mitter, C. & D. R. Brooks. 1983. Phylogenetic aspects of coevolution. *In*: D. J. Futuyma & M. Slatkin (eds.), *Coevolution*. Sinauer Assoc., Inc., Massachusetts.

Moler, P. E. 1985. A new species of frog (Ranidae: *Rana*) from northwestern Florida. *Copeia* 1985: 379–383.

Mooney, H. A. & S. L. Gulmon. 1983. The determinants of plant productivity—natural versus man-modified communities. *In*: H. A. Mooney & M. Godron (eds.) *Disturbance and ecosystems.* Ecological Studies 44, Springer-Verlag, New York.

Myers, N. 1979. *The sinking ark.* Pergamon Press, New York. 307 p.

Ono, R. D., J. D. Williams & A. Wagner. 1983. *Vanishing fishes of North America.* Stone Wall Press, Washington, DC. 257 p.

Pirsig, R. M. 1974. *Zen and the art of motorcycle maintenance.* Morrow Quill, New York. 412 p.

Qi, Y., C. W. Reimer & R. K. Mahoney. 1982. Taxonomic studies of the genus *Hydrosera.* I. Comparative morphology of *H. triquetra* Wallach and *H. whampoensis* (Schwartz) Derby, with ecological remarks. *Proc. 7th Int. Diatom Symposium*

Rathke, B. J. 1984. Patterns of flowering phenologies: testability and causal inference. *In*: D. R. Strong et al. (eds.) *Ecological communities: conceptual issues and the evidence.* Princeton Univ. Press, New Jersey.

Ricklefs, R. E. 1973. *Ecology.* Chiron Press, Massachusetts. 861 p.

Selander, R. K. 1969. The ecological aspects of the systematics of animals. *In: Systematic Biology. Proceedings of an international conference.* National Acad. Sci. Publ. 1692, Washington, DC.

Sibley, C. G. & J. E. Ahlquist. In press. The relationships of some groups of African birds, based on comparisons of the genetic material, DNA. *Proc. Symp. African Vertebrates, Bonner Zool. Beitrage.*

Simberloff, D. 1983. Sizes of coexisting species. *In*: D. J. Futuyma & M. Slatkin (eds.), *Coevolution.* Sinauer Assoc., Inc., Massachusetts.

Simberloff, D. 1984. Properties of coexisting bird species in two archipelagoes. *In*: D. R. Strong et al. (eds.) *Ecological communities: Conceptual issues and the evidence.* Princeton Univ. Press, New Jersey.

Simpson, G. G. 1961. *Principles of animal taxonomy.* Columbia Univ. Press, New York. 247 p.

Slater, J. A. 1981. The pursuit of the smallest game. A look at systematic entomology from the 21st century. *Trans. Am. Ent. Soc.* 107: 149–162.

Smith, M. L. 1981. Late Cenozoic fishes in the warm deserts of North America: a reinterpretation of desert adaptations. *In*: R. J. Naiman & D. L. Soltz (eds.), *The biology of fishes in North American deserts.* Wiley Interscience Series, New York.

Soule, M. E. & B. A. Wilcox (eds.). 1980. *Conservation biology.* Sinauer Assoc., Massachusetts. 395 p.

Starnes, L. B. & A. E. Bogan. 1982. Unionid mollusca (Bivalvia) from Little South Fork Cumberland River, with ecological and nomenclatural notes. *Brimleyana* 8: 101–119.

Stearns, S. C. 1984. The effects of size and phylogeny on patterns of covariation in life history traits of lizards and snakes. *Am. Nat.* 123: 56–72.

Theriot, E. & E. F. Stoermer. 1982. Principal component analysis of character variation in *Stephanodiscus niagarae* Ehrenb.: Morphological variation related to lake trophic status. *Proc. 7th Int. Diatom Symposium* 1982: 97–111.

Tilman, D. 1982. *Resource competition and community structure.* Monographs in Population Biology No. 17, Princeton Univ. Press, New Jersey. 296 p.

Walker, T. J. 1964. Cryptic species among sound-producing ensiferan Orthoptera (Gryllidae and Tettigoniidae). *Q. Rev. Biol.* 39: 345–355.

Whitcomb, R. F., C. S. Robbins, J. F. Lynch, B. L. Whitcomb, M. K. Klimkiewicz & D. Bystrak. 1981. Effects of forest fragmentation on avifauna of the eastern deciduous forest. *In*: R. L. Burgess & D. M. Sharpe (eds.) *Forest island dynamics in man-dominated landscapes.* Ecological Studies No. 41, Springer-Verlag, New York.

Whitmore, T. C.(ed.). 1973. *Treeflora of Malaya.* Vol. 2. Forest Dept., Ministry of Primary Industries, Kuala Lumpur, Malaysia, 106 p.

Wilbur, H. M. & J. Travis. 1984. An experimental approach to understanding pattern in natural communities. *In*: D. R. Strong et al. (eds.) *Ecological communities: conceptual issues and the evidence.* Princeton Univ. Press, New Jersey.

Wilcox, B. A. 1980. Insular ecology and conservation. *In*: M. E. Soule & B. A. Wilcox (eds.), *Conservation biology.* Sinauer Assoc., Massachusetts.

Wiley, E. O. 1981. *Phylogenetics. The theory and practice of phylogenetic systematics.* Wiley-Interscience, New York. 439 p.

Diversity, Germplasm and Natural Resources

Christine Schonewald-Cox
National Park Service, U.C.–Davis

Abstract: The conservation of genetic diversity, germplasm, and natural resources depends upon knowledge of natural fauna and flora obtained through a biological survey. The author discusses the intrinsic values of genetic diversity, germplasm, and natural resources in increasing the effectiveness of conservation and making it possible for us to both slow extinction rates and detect extinctions as they occur or are pending. She discusses the need for a biological survey and the valuc of establishing, monitoring, and updating a data base in the context of these issues. She reviews recommendations made by international conservation organizations that also recognize the essential importance of biological surveys.
Keywords: Genetic Diversity, Germplasm, Natural Resources, Biological Survey.

INTRODUCTION

The conservation and management of genetic and biological diversity, germplasm, and natural resources will not proceed systematically without extensive knowledge of the natural biota. "NABIS," the proposed National Biological Survey, has the potential to provide us with this knowledge, which is greatly needed by all communities concerned with conservation and human welfare and survival. In this review, I will first address myself to the intrinsic value of such things as diversity, germplasm, and natural resources in promoting the importance of a national biological survey. Human survival and the conservation of diversity, germplasm, and natural resources depend upon knowledge of the status, variety, and distribution of, including changes in, our biota. It is this knowledge that forms the basis upon which are developed ideas and discoveries that support both our survival and the continued abundance of other species. It is the lack of such knowledge that will undermine even our own existence. Thus, why we need a national survey, such as NABIS, to provide this knowledge is the second issue I will address. And, finally, I will bring attention to current and compatible efforts that relate to a national biological survey.

INTRINSIC VALUE

Terms and Definitions

The terms "diversity" (whether genetic or biological), "germplasm," and "natural resources" carry sometimes contradictory meanings, depending upon the orientation of the user and the context in which they are used. "Natural resources", for example, is used as a category of responsibility by the National Park Service. This category includes the variety of flora, fauna, and non-living natural and scenic qualities protected by a park, such as geologic, fossil, and visibility attributes of the park, but does not include historic or cultural structures or artifacts. In the general usage of the term, prevalent in the literature, however, "natural resources" carries an exploitative connotation (forest lumber, minerals, soils, or water), which is contradictory to the definition used by the National Park Service. Similar usage problems exist with the terms "genetic (or biological) diversity" and "germplasm." Therefore, I'll first introduce the subjects with definitions.

Genetic Diversity

Genetic diversity can be defined as variation in genetic composition: variation within and between individuals, populations, species, and higher taxa are implicit (also categorized under the term "biological diversity").

Every individual blade in a patch of grass may be genetically identical, or nearly so, but if one examines different patches of the same grass, each may prove to be genetically unique. At the same time, a genetic analysis of a different species may reveal that all its populations are genetically similar but that individuals within the populations vary greatly. If we wish to conserve either of these types of species, we may do so by utilizing knowledge of their geographic and temporal apportionments of genetic, including phenotypic, diversity. But without studies throughout species' ranges, these types of information will be unavailable. Moreover, most species, with several, possibly rare and unique, genotypic or phenotypic traits, may remain incompletely described without a systematic survey of North American habitats. Genetic diversity also manifests itself in the ecological interactions of species. This can be observed in unique ecological relationships that have developed, such as between keystone species and their dependent species. Once again, only a systematic survey can hope to reveal the complexity and biogeographic extent of these relationships.

In the field of evolutionary biology, considerable attention has been given to how genetic diversity is manifested by polymorphisms, proportions of heterozygosity, and linkage groups. For example, stresses imposed on portions of animal and plant populations can significantly alter the genetic structure of individuals locally, sometimes favoring rare alleles in heterozygous conditions, the presence of which, under normal circumstances, could decrease fitness of individuals [(Liu and Godt, 1983; as with responses of rats to warfarin and concomitant increase in minimum requirements of vitamin K as described by Bishop (1981) and Bishop and Hartley (1976)]. The significance of this lies in the flexibility of responses by the genome to the environment provided by variability (polymorphisms, heterozygosity, and differing linkage groups). The accumulation of such knowledge offers interesting possibilities for understanding how species respond to new environments and how evolution functions in species' adaptations to stress.

Germplasm

Germplasm can be defined as genetic material in the cell. Current use suggests that germplasm refers primarily to genetic material that is the template for new agricultural, industrial or medicinal developments, as well as providing dried or frozen material for genetic and systematic analysis and preserved cells, embryos, seeds, etc., of ancestral types and close relatives of domesticates.

Whether for studies in systematics and evolutionary biology or for preservation of semen, embryos, eggs, pollen, seeds, or cell lines and tissues, the preservation of germplasm constitutes a valuable investment in the advancement of technology for ensuing decades and beyond.

Germplasm is not only a resource for future use, but is already being used extensively at present. In systematic biology we sometimes depend on frozen tissues in the electrophoretic analysis of population characteristics or in DNA hybridization, for example, for determining the common ancestries of similar taxa and to determine the probable evolutionary distance between them. Most of the work dependent upon germplasm has come of age in the 70's and 80's. As Dessauer and Hafner (1984) point out, this increased interest is evidenced by the preponderance of papers on molecular analysis in systematic journals.

Zoological parks and botanical gardens have taken it upon themselves to maintain species *ex situ* that are nearly extinct *in situ.* Aside from the manipulation of breeding systems, diet, and microclimate, new laboratory techniques are being developed that accelerate population growth. These techniques to which I refer are: frozen storage of sperm, embryos, eggs (this one with little success so far), and pollen and seed banks. Frozen sperm can be used to increase the effective population size dramatically by either decreasing generation time or balancing individual contributions to reproduction and sex ratios, for example. Embryos from close taxonomic relations transplanted from one female to another can also be used to increase the effective population size by increasing fecundity and birth rate. Similarly, pollen and seed banks function to increase fecundity and reduce generation time, and, as with animals, retain genetic options for increasing diversity in selected breeding populations.

The function of cell line and tissue storage is less immediately obvious. Traits kept in cell lines from disease-tolerant or toxin-tolerant animal and plant species can be kept for future technology to reproduce and share with non-tolerant individuals, agricultural varieties, or endangered species. Organs and tissue lines of vanishing species may be kept for future investigation of their potential contributions, not only to the parent species, but also to human survival—new medicines, food crops, or detoxifying agents, for example. Opportunities for advancement are limited only in our conception and imagination of what gifts the storage of germplasm can offer the future.

Natural Resources

Natural resources can be defined as capacities or materials supplied by nature. Current use suggests those materials that have value in sustaining life as well as lifestyle for humans. This term is sometimes equated with genetic and scenic, excluding cultural, diversity combined.

Most natural resources that have been extensively described or inventoried currently support strong financial interests and production. This is the case for most major crops, livestock, minerals, and waterways, for example. But what of those resources for which we have not yet discovered uses? Or, what of those needs for which we have not yet found the resources? It seems we know all too little about what our continent has to offer in the realm of future needs and possibilities. And, to beg the question, are any of the United States' biological resources non-renewable?

I believe that exploited species for which recruitment in the population is consistently less than the consumption rate are, essentially, non-renewable. And, with reference to recruitment, I am speaking not only in terms of absolute numbers of individuals but in terms of the proportion of natural genetic variability renewed with every generation. This will determine, in part, the evolutionary potential of the population or species. In such a case, probably including most of the non-domesticated species, individuals that are directly exploited (perhaps excluding mallard ducks and whitetailed deer) qualify as non-renewable natural resources.

Humans have historically not gone out of their way to protect life that does not relate to human survival or pleasure. But, fortunately, the public definition of pleasure is rapidly changing. A new segment of our population has come to value habitats and species from which they may never realize any tangible benefits, other than the pleasure of seeing wildlife through mass media. The wildlife seen through media have become an important natural resource, in the public's mind, and will only increase in importance with time.

THE NEED FOR A BIOLOGICAL SURVEY

So, what does a biological survey offer toward increasing our knowledge of biological diversity, germplasm, and natural resources? First, it results in a data base. Second, it involves monitoring and updating the data base. And, third (most importantly) it furnishes the knowledge we require. Inevitably, such a project will increase cooperation between science, agriculture, industry, medicine, recreation, and other groups to build a data base and plan a future for the biosphere and ourselves, collectively.

Building the Data Base

What information should the data base gather during its building phase in terms of diversity, germplasm, and natural resources? The data base that results from a national biological survey has the potential to contain documentation of genotypes and phenotypes of species at specified places and times. Multiple documentations over time in the same place can offer us an ability to track changes in habitats and species that occur both "randomly" and in response to environmental changes. Similarly, multiple documentations in different locations offer the opportunity of measuring differences that exist between populations and between habitats. A biological survey will forcibly document the distributions and abundance of wild relatives of domesticated species and of habitats with the greatest potential for housing biological diversity and maintaining valued natural resources. The data base provided by a biological survey can provide the baseline data against which changes will be measured in the future.

The Value of Establishing, Monitoring, and Updating the Data Base

Analyses of intervals in both time and space offer the opportunity to detect new genotypes, populations, species, or ecological manifestations; to detect changing conditions (e.g., health, vigor, etc.); and to resolve which changes among these are humanly induced; as well as to discern which events or variables are the possible agents of change and causes of spatial and temporal difference.

Furnishing Knowledge

The potential knowledge furnished by the data base is vast and beyond the scope that any of us can presently predict. However, in the realm of conservation and management of diversity, germplasm, and natural resources, the survey's product data base has some predictably valuable uses.

The data base can be used for developing new conservation biology, medical, agricultural, and wildlife management theory as well as improved techniques and practices. A data base such as this will facilitate the discovery of new medicines, foods, substitutes, or life- and lifestyle-sustaining resources.

Equally important to these products is the monitoring of the speed at which species expand their ranges and diversify. This is a subject that is poorly known despite numerous ongoing investigations into evolutionary processes. Some species change discernably within our lifetimes while other changes seem to require geological time. The data base provided by the biological survey offers the opportunity to measure the rates of change as well as the relative rates of anthropogenic versus natural extinction (when these can be distinguished from each other).

The systematized knowledge of the existence of the diversity of species, germplasm, and natural resources will help us prescribe the magnitude of their protection needs. In addition, this knowledge will provide the working matter needed to develop sound conservation legislation and management mandates necessary to bring about this protection.

Because of its "national-scale" of activity, the biological survey has a potential to increase public awareness and thereby stimulate public conscience relative to maintaining diversity, germplasm, and natural resources in the U.S. This is not only achieved by the publicity and television programs that arise out of the effort but also by the fact that the undertaking of the biological survey affects employment and public exposure. The public and involved working force come into contact with the non-consumptive value of wildlife (including plants) as well as the consumptive uses of renewable resources and maintenance of germplasm that are equally crucial to human survival.

Refining the Data Base

The availability of automated data processing, particularly microcomputer networking systems, offers a host of opportunities to maintain, modify, and utilize the data base collected in the biological survey. Refinement of the data base would be undertaken to incorporate changes in scientific thinking about documentation, changes in species and population relationships, and changes in the types of agricultural varieties or pests. The refinement is the logical ongoing product of monitoring and use of the data base. The fact that a survey is conducted comes with at least partial assurance that improvements and modification through use will inevitably be made. That in itself is a positive outcome of such a survey effort.

Cooperation

It is the building and use of a broad and updated data base that brings together scientists and their institutions to cooperate on large interdisciplinary projects. Incorporated within this framework of cooperation will be the increased role that systematics, biogeography, and collections housing institutions (e.g., museums) could play in maintaining the records and vouchers for the survey.

Because other countries already have, or are currently developing, their national surveys, opportunities will only increase for international cooperation to improve understanding of natural ecological processes and anthropogenic impacts. Cooperation, whether it is at the private, public, institutional, or international levels, will occur on the basis of interdependencies (Salwasser et al., in press) and for the facilitation of information and talent exchange in conservation and management of diversity, germplasm, and natural resources.

WHAT ARE CURRENT NATIONAL TRENDS THAT RELATE TO A NATIONAL BIOLOGICAL SURVEY?

Foremost, modern conservation and economic philosophies beg for a biological survey, whether or not they know it. The conservation community needs an enlightened approach to an expanding problem—extinction. The economic community desires to survive and thus needs to keep up its own "food" supply. Dasmann (1973a and 1973b) and later Udvardy (1975) have endeavored to classify the biological realms and biomes of the earth to allow systematic searches and assessments to be made of their diversities of organisms. Recently Hayden et al. (1984) have developed a parallel biological classification for the major bodies of water. The Nature Conservancy and associated Natural Heritage Programs have attempted to set aside "unique" communities based on what knowledge they have been able to gather on the distribution and uniqueness of each species or community. The International Program on Man and the Biosphere has endeavored to take maps of large-scale biogeographic biomes and provinces and high resolution vegetation maps (such as Kuchler, 1966) and include a sample of each habitat type in its biosphere reserve network. It is hoped by MAB that biosphere reserves will contain representative diversity of the biomes or habitats in which they are established. While selection of sites for biosphere reserves is hardly done blindly, MAB has recognized the difficulty of finding biological survey-like data to systematize their selection efforts.

As the State of the Environment Report (1984) states, "Historically, the greatest concern for, and therefore the most information about, wildlife has focused on a relatively small group of species—those commercially valuable or taken for sport (for example salmon and deer), those with special aesthetic appeal to amateur naturalists (birds, butterflies), or those especially awesome or attractive (whales, owls, grizzly bears). Therefore, most readily available data on wildlife are on vertebrates, particularly mammals and birds, and especially game species." The endangered and threatened species lists developed by many of our states contain considerable numbers of species that have not yet been added to the national list. This would probably not be the case if we had a national data base from which the national listing office could obtain data on distributions and abundance. Interestingly, the majority of species that are listed tend to be the scientifically better known species or taxa. While a high priority has been given to looking at invertebrates (and a few are included on the national list) does it mean that only a few invertebrates are actually threatened with extinction? How can we begin to answer this when we have not determined what taxa naturally exist here in the U.S., and at what natural (and periodic) abundance levels?

Ongoing surveys such as those conducted in a few states, the National Wetlands Survey, and federal surveys of barrier islands, are preliminary survey projects that really should be part of the larger coordinated effort embodied in a national biological survey. The establishment of biosphere reserves, the selection of protected habitats, and the allotment of areas for mutiple use, such as for national forests or game refuges, should follow from the results of a survey rather than be the subject of its justification.

CONCLUSIONS

As concerns species and higher taxa, approximately 50% of the Hawaiian bird species are thought to have been extirpated by humans and related causes. We are fortunate in having caves in Hawaii where bones have been discovered that help us to reconstruct the previous avifauna of these islands (Olson and James, 1982; James and Olson, 1983). Such finds act similarly to the museum collections in which surveys produce specimens that vouch for previous conditions.

In his address at the first International Conference on Biological Diversity (1982), Khoshoo recalled, "There was a physician living 3000 years ago who was asked by his teacher to find a plant which was useless. He returned after 10 years, saying that there was no such plant." This decidedly eastern parable bears upon the attitude of the U.S. towards its native resources. Can the U.S. acknowledge ignorance—that there is more to learn about one's environment beyond a prescribed achievement than the development of economic and agricultural superiority based upon approximately 1,000 *imported* economic species? Are our flora and fauna so useless or less attractive that they are not worth looking at? At least 150,000 people represented by scientific organizations led by AAAS think our resources are worth tracking. As Kosztarab (1984) pointed out, the proposal to establish a national survey, to fund the basic taxonomic research, and to produce the manuals, catalogs, atlases, and classification systems is already before Congress. What we seem to need is for someone *not* to be afraid to make a "profitable" decision.

RECOMMENDATIONS

The ASC, on an only geographically smaller scale, is automatically joined by a strong unity of purpose shown in recommendations made by the International Union for the Conservation of Nature, World Wildlife Fund, international MAB, UNEP, IBPGR, FAO, UNDP, The 1st International Conference on Biological Diversity (U.S. Department of State, 1982), and The World Conservation Strategy (IUCN, UNEP, and WWF, 1980), to name a few organizations and efforts which recommend that the world's resources (including the vast genetic diversity) be documented. As concerns resources within the U.S., I would prefer to iterate the collection of excellent recommendations already made by these groups adopted to a national scale:

1. Establish a national register of genetic resources.
2. Develop knowledge of the status of existing plant and animal resources in the U.S.
3. On the basis of the abundance and distribution of natural diversity, establish voucher collections and depositories that systematically preserve animal and plant specimens and germplasm, including tissue samples and biological reagents (such as antisera). Do this on the basis of the function of collections: A) specialized research, B) major reference, and C) national resources.
4. Inventory ecosystems and accelerate efforts to identify and protect areas that represent unique ecosystems and which have high species diversity that remain unprotected.
5. Monitor unique ecosystems by examining environmental conditions and indicator species.
6. Any survey team should include specialists in organismic biology, systematics, comparative taxonomy, and ecology for the full range of organisms that enter into the survey.
7. The U.S. should establish a U.S. interagency task force on biological diversity to develop comprehensive, long-term goals and strategies to inventory and maintain biological diversity.
8. The function of the survey should be institutionalized at the federal and state levels.
9. Coordinate the inventory with Canada and Mexico and evaluate all current international and within-U.S. efforts to inventory both wild and introduced species and varictics.
10. Refine our ecosystem classification system on the basis of the results of the national biological survey to guidc selection of areas to be managed for multiple use or for protection of native biological diversity and introduced diversity threatened with extinction in its native habitat.

ACKNOWLEDGEMENTS

I would like to thank Thomas Gilbert (formerly of the National Park Service), Arthur Allen (National Park Service), John Dennis (National Park Service), Regional Curators (National Park Service), Jonathan Bayless (National Park Service) and Jacqueline Schonewald (California Academy of Sciences, retired) for their helpful discussions on the topics of biological inventory and voucher collections, and for having stimulated my interest in this topic. Thanks also to Richard Baker (National Park Service) for his editorial assistance. The views expressed in the paper are strictly those of the author and do not necessarily reflect the opinions nor policies of the National Park Service.

LITERATURE CITED

Bishop, J. A. 1981. A neo-Darwinian approach to resistance: examples from mammals. *In*: J. A. Bishop and L. M. Cook (eds.). *Genetic consequences of man-made changes.* Academic Press, London.

Bishop, J. A. & D. J. Hartley. 1976. The size and age structure of rural populations of *Rattus* *norvegicus* containing individuals resistant to the anticoagulant poison Warfarin. *Jr. of Animal Ecol.* 45: 623–647.

Dasmann, R. F. 1973a. *A system for defining and classifying natural regions for purposes of conservation.* I.U.C.N. Occasional Paper No. 7, Morges, Switzerland.

Dasmann, R. F. 1973b. *Biotic provinces of the world.* I.U.C.N. Occasional Paper No. 9, Morges, Switzerland.

Dessauer, H. C. & Mark S. Hafner (eds.). 1984. *Collection of frozen tissues: value, management, field and laboratory procedures, and directory of existing collections.* Association of Systematics Collections, Lawrence, Kansas. 73 p.

Hayden, B. P., G. C. Ray & R. Dolan. 1984. Classification of coastal and marine environments. *Environ. Conserv.* 11(3): 199–207.

I.U.C.N. 1980. *World conservation strategy, living resource conservation for sustained development.* I.U.C.N., U.N.E.P., and W.W.F.

James, H. F. & S. L. Olson. 1983. Flightless birds. *Natural History* 92(9): 30–40.

Khoshoo, T. N. 1982. Conservation of biological diversity: the Indian experience. *In:* U.S. Department of State (ed). *Proceedings of the U.S. Strategy Conference on Biological Diversity, November 16–18, 1981.* U.S. Department of State Publication 9262 (International Organization and Conference Series 300) B.I.O.A. 126 p.

Kosztarab, M. 1984. Editorial: A biological survey of the United States. *Science* 223(4635): 443.

Kuchler, A. W. 1966. Potential natural vegetation. *In: The national atlas of the United States of America.* U.S.D.I.-U.S.G.S.

Liu, E. H. & M. J. Godt. 1983. The differentiation of populations over short distances. *In*: Schonewald-Cox, C. M., S. M. Chambers, B. MacBryde and L. Thomas (eds.) *Genetics and conservation.* Benjamin Cummings, Menlo Park, California. 722 p.

Olson, S. L. & H. F. James. 1982. Fossil birds from Hawaiian islands: evidence for wholesale extinction by man before western contact. *Science* 217: 633–635.

Risser, P. G. & K. D. Cornelison. 1979. *Man and the Biosphere: U.S. Information Synthesis Project MAB-8 Biosphere Reserves.* Oklahoma Biological Survey. Norman, Oklahoma.

Salwasser, H., C. Schonewald-Cox & R. J. Baker. *In press.* The role of interagency cooperation in managing for viable populations. *In*: Soule, M. E. (ed.) *Viable populations.* Cambridge University Press. Cambridge, Massachusetts.

The Conservation Foundation. 1984. *State of the environment: an assessment at mid-decade.* The Conservation Foundation. Washington, DC.

Udvardy, M. D. F. 1975. *A classification of the biogeographic provinces of the world.* I.U.C.N. Occasional Paper No. 18.

U.S. Department of State. 1982. *Proceedings of the U.S. Strategy Conference on Biological Diversity, November 16–18, 1981.* Department of State Publication 9262 (International Organization and Conference Series 300) B.I.O.A. 126 p.

The Role of a National Biological Survey in Environmental Protection

Allan Hirsch
U.S. Environmental Protection Agency

Abstract: A national biological survey could make a major contribution to environmental protection and management in the U. S. To be most useful for that purpose, the survey should include information on population biology and ecology as well as on taxonomy. The survey should also focus on means of making existing biological information available and on synthesizing such information to address environmental management needs. Strong emphasis should be placed on development and use of modern information system technology.
Keywords: Biological Survey, Environmental Protection, Environmental Impact Assessment, Ecosystem Classification and Inventory.

INTRODUCTION

A national biological survey, well conceived and effectively managed, could make a major contribution to environmental protection and management in the U.S.

There can be little doubt about the importance of the availability of adequate information about flora and fauna in addressing such issues as toxic chemical pollution, acid deposition, land disturbance, pesticide resistance, and the environmental impacts of introduced plants and animals, including new genotypes from the emerging biotechnology industry. Vast amounts of biological data have already been collected, sometimes in an attempt to address these very problems. However, the data are widely scattered, often not accessible or comparable, and perhaps most important, generally not organized in formats relevant to environmental decision making. There are also major information gaps that can be filled only by additional research and survey work. Therefore, the current interest in establishing a national survey is highly encouraging.

Special efforts will be needed in design and execution of the survey if it is to serve the specific needs of environmental protection, as well as meet other goals such as support of basic biosystematics research. My remarks will focus on the information needed to support applied environmental protection and management and on recommendations concerning how those needs might be addressed.

TAXONOMIC INFORMATION

A biological survey can be described as an organized effort to characterize the biota of a nation or region. Typically, biological surveys have been grounded primarily in basic taxonomic work on the classification and distribution of biota, resulting in the production of monographs, identification keys, museum collections and, increasingly, computerized information bases.

There is an important need to fill gaps in taxonomic information in support of environmental research and monitoring. For example, reduction in species diversity may provide subtle but important early warnings of long term environmental trends. For some groups, such as microorganisms or invertebrates, basic limitations in taxonomic information may impede recognition of such symptoms. Water quality evaluations are made through construction of diversity indices or through use of various indicator organisms. Species level identification may be important, as different species within the same genera can exhibit a wide range of pollution tolerances. For example, insects and other macroinvertebrates have long been used as indicators of water quality. However, in many groups of aquatic insects, identification of the immature stages usually collected in stream surveys cannot be made below the generic level. Improved taxonomic information could improve the sensitivity and reliability of biological water quality evaluations (Resh and Unzicker, 1975).

Taxonomic information of the kind likely to be provided by a biological survey is often directly and immediately relevant to the operating requirements of applied environmental protection as well as meeting more basic research needs. I will give several examples.

One of the principal environmental impacts of electric generating power plants is destruction of ichthyoplankton by impingement and entrainment in intake water systems. A central issue in siting such plants or in determining required controls on plant design and operation is the presence or absence of large concentrations of ichthyoplankton, particularly species of commercial or recreational importance. Environmental consultants and others conducting impact studies were having difficulty in identifying ichthyoplankton and determining what species were involved. The U.S. Fish and Wildlife Service (FWS) prepared a massive atlas of egg, larvae and juvenile stages of fishes found on the mid-Atlantic coast to help meet this need (Jones et al., 1978).

As I have already indicated, benthic invertebrates are important in the assessment of water quality conditions. The U.S. Environmental Protection Agency (EPA) sponsored preparation of a series of keys to freshwater invertebrates to facilitate pollution assessment (U.S. EPA, 1973).

The national Clean Water Act provides for the protection of wetlands. An important issue in determining whether this regulatory jurisdiction applies at a given site is the determination of wetland boundaries. Hydrophytic vegetation and hydric soils are key wetland indicators. A Federal interagency group has been developing comprehensive regional lists of hydrophytes and hydric soils, and FWS is developing a computerized National Wetlands Plant database.

These rather obvious examples indicate that operating agencies on the environmental "firing line" sometimes need basic taxonomic information in support of their missions. The importance of taxonomic studies of neglected taxa and regions should not be minimized; that should come as no surprise to this audience.

However, in seeking support or sponsorship for a national biological survey from agencies responsible for environmental protection, it is important from the outset not to overstate the need for taxonomic data. In my opinion, in applied environmental management, the most important need is for other kinds of biological information. An understanding of environmental impact requires the ability to identify the flora and fauna of the area under study, but from the standpoint of practical environmental management, often this understanding need not be a comprehensive ability to identify all forms.

To cite a case in point, a very useful service the survey could perform would be provision of information to support preparation of environmental impact statements. Many millions of dollars have been spent in biological baseline surveys for such statements, but the resultant information has often been lost after initial use. The survey could reduce both redundant data collection and subsequent data loss. Yet an aspect of environmental impact statement preparation that has been widely criticized both in terms of costs and relevance to decision making has been inclusion of exhaustive species lists. Other kinds of biological information may be more useful in assessing environmental impacts and in identifying appropriate protective measures. Effective impact assessment is evolving towards a more selective evaluation of various ecosystem components or species of special concern (Beanlands and Duinker, 1985).

Therefore, to play an important role in meeting environmental management needs, the national biological survey must go beyond strictly taxonomic considerations to include information on the distribution and population patterns and trends of important species and on ecosystems.

POPULATION BIOLOGY

The need for information on population structure, status, and trends is most evident in the case of managed species, where an understanding of the population structure is essential to determine harvest rates and other management actions. Somewhat the same considerations apply with respect to environmental protection, as well. We may be able to forecast the impacts of a development on organisms or their habitat, but a broader context is needed to determine the significance of those impacts and to make social judgements concerning their acceptability when choices must be made between development and environmental values. We must be able to address such questions as what percentage of the species population or its range is being affected? Is the impact stressing a species or population already in decline?

For example, in connection with development of the oil and natural gas resources of the Outer Continental Shelf, concern has been considerable about potential impacts of oil spills and other activities on marine bird and mammal populations. This concern has led to the sponsorship of extensive studies and surveys of the distribution and habits of marine bird and mammal populations by the U.S. Department of the Interior, which is responsible for the leasing program. In addition to contributing greatly to our general knowledge of the species concerned, these studies have resulted in extensive information that has been used to exclude certain particularly sensitive areas from leasing and to modify certain aspects of the development program, such as limiting exploratory drilling in the vicinity of important bird colonies during the nesting period.

At first glance, maintenance of population information may appear to be a Herculean task, involving not only inventories but also periodic monitoring to assess status and trends. Yet a significant start could be made by providing reports that consolidate and interpret existing data to provide a synoptic or historic view. Probably the richest source of population information is data on managed species collected by fish and wildlife agencies. Breeding bird surveys and Audubon Christmas bird counts provide information on the status and trends of avifauna. Other sources are found in various individual research studies.

Surveys of habitat loss also provide important correlative information. The effects of forest habitat fragmentation are reflected in reductions in the population of warblers and other passerine birds. A report on wetlands status and trends, compiled by the FWS National Wetlands Inventory (Tiner, 1984), which interprets past and current aerial photography to determine rates of wetlands loss, could be used in concert with population trend information on waterfowl and other fauna. Habitat inventories often have the advantage of being easier to conduct than direct population surveys; for many types of animals we still lack reliable and cost-effective methods to inventory and monitor population numbers on a regional or national scale.

Systematic interpretation of existing information, perhaps on a national basis, would be an important first step. Initial efforts of a national biological survey to compile information on population distribution and trends would, of necessity, focus on high interest species. Over time, however, a broader picture might emerge that would provide realistic assessments of trends in loss of genetic diversity and provide early warning concerning species that ultimately will become candidates for threatened and endangered species lists.

BIOLOGICAL MONITORING

Maintaining information on the status and trends of various species is closely related to the issue of biological monitoring. Monitoring to assess environmental conditions is one of the most useful applications of biological information in environmental management.

An important application of biological monitoring is the use of organisms as accumulators of contaminants. Various organisms used for this purpose include honeybees, earthworms, birds, and fish. FWS studies of contaminants in fish and in such ubiquitous birds as starlings and ducks have provided valuable information on environmental build-up and decay of PCB and DDT. EPA developed the protocol for a marine monitoring system called "Mussel Watch", in which mussel tissue from various estuaries was analyzed to assess build-up of contaminants in the marine environment. The monitoring of fish tissue contaminated with metals and chlorinated hydrocarbons is routinely used to determine the need for public health limitations on fish consumption. A related aspect that has received increased attention, and which might be relevant to those elements of a national biological survey associated with museum collections, is establishment of tissue banks to provide a basis for future comparison of environmental conditions.

A second aspect of biological monitoring is use of organisms themselves as indicators of environmental quality. There are many examples. Evaluation of benthic invertebrate and diatom populations has been a long-standing tool in water quality assessments. Absence of lichens can be an indicator of air pollution. Recent reports of failures of salamanders to spawn have been interpreted as an early warning of possible increases in acid deposition in the Rocky Mountain area. This category of monitoring can be more difficult to evaluate because natural spatial and temporal variability make both detection and interpretation of human-induced changes difficult, whereas measuring anthropogenic compounds may be somewhat more straightforward.

The FWS and EPA recently completed a joint fisheries survey, described as the first statistically designed national survey of the status of waters, fish communities, and limiting factors affecting those communities (Judy et al., 1984). A statistical sample of river segments was assessed by use of questionnaires and existing data bases. The findings, while preliminary, are important because they serve as a general indicator or surrogate for the quality status of the nation's waters and because they also provide important information on the status of the fishery resource. This survey used information on sport, threatened and endangered fish species, and other species of special concern as indicators of biological status. There also have been other efforts to develop more comprehensive fish community analyses for use in environmental assessment (Karr, 1981).

An intriguing linkage of both aspects of biomonitoring was suggested in a report of the National Science Foundation on long term ecological measurements (National Science Foundation, 1977). That report identified seabird populations as potentially important indicators of marine environmental quality. Marine birds are long-lived and, although widely dispersed much of the year, are highly concentrated during the nesting season, so reasonably accurate statistical sampling can be conducted. Because these birds are high in the food chain, they are potential accumulators of contaminants as well as integrators of ocean ecosystem conditions. To detect wide scale environmental changes in the ocean, it might be feasible to design long term sampling programs that combine reliable monitoring of nesting areas through aerial photography, species composition studies, and sampling of tissue and eggs for contaminants. Thus, seabirds might be used as an early warning system to detect impacts of ocean contamination.

The scope and scale of biological monitoring can vary widely, as can its fundamental purpose. Biological monitoring can be used to measure site-specific changes, as in the case of benthic organisms downstream from an effluent discharge. Currently EPA is using biomonitoring to measure the toxicity of complex industrial effluents and to assure that industrial plants comply with pollution control requirements. There is probably little role for a national biological survey in such site-specific applications.

Increasingly, however, environmental protection programs involve evaluation of problems of wide-scale regional or even global scope. Carefully designed long-term biological monitoring can help detect wide-scale biosphere changes associated with such environmental problems as acid deposition, climate warmings from the "greenhouse effect", ozone reductions by chlorofluorocarbon discharges, and long-range atmospheric transport of toxic contaminants.

The United Nations Environment Program has initiated a Global Environmental Monitoring Program, which incorporates ecological components, and the National Science Foundation has sponsored development of a long-term ecosystems monitoring network on carefully selected experimental ecological reserves to encourage a systematic approach. UNESCO's Man and the Biosphere program also sponsors long term monitoring in biosphere reserves. Although a number of such programs are already underway, progress is spotty. A national biological survey could help provide standardization, comparability, and continuity in those aspects of biological monitoring that involve a long term commitment and wide scale regional or global networks.

ECOSYSTEM CLASSIFICATION AND INVENTORY

Thus far, I have focused largely on information on plant and animal species. However, much of the thrust of environmental protection is to address the full range of impacts that occur at a place or a class of places (e.g., salt marshes, arctic tundra). The geographic scope of concern may be narrowly site-specific or very broad in scale. We may be evaluating impacts of hazardous wastes in the immediate vicinity of a disposal site, addressing the effects of a nuclear electric power generating plant on an estuary, or assessing the impacts of acid deposition in the Northeastern United States. In each case, we are concerned with effects on the ecosystem as well as on individual components. A national survey could provide important information on the distribution of ecosystem types, analogous to its role in the systematic compilation of information on taxa.

Ecological inventories are assuming increasing importance in environmental management. For example, the FWS National Wetlands Inventory, which surveys and maps the distribution of wetlands nationwide, provides essential information for use in wetland protection programs. EPA is currently conducting a National Lakes Survey to assess trends in acidification as part of a nationwide acid deposition research program. Inventories of the distribution of ecological communities collected by the Nature Conservancy and its State Natural Heritage Programs provide an important guide to the Conservancy's acquisition and preservation programs.

Ecological classification systems are integral to an inventory effort. Such systems help us organize knowledge by simplifying complex interrelationships to identify geographic areas and ecosystems with similar properties. Classification systems provide the structure for designing inventory programs and aggregating the resultant masses of data.

Because similar ecological units should respond in a like manner to the same environmental stresses or management remedies, classification systems also increase our ability to generalize, to extrapolate research results from one area to another, and to transfer management experience. Thus, ecological classification systems can be a powerful tool in environmental management.

There are various approaches to ecosystem classification, and I will not discuss them in detail here. However, any system designed for a national biological survey should probably be a primary system based on biosphere components such as climate, soils, and vegetation. Such a system could also serve as a foundation from which to define land classes in terms of various environmental or resource management attributes and potentials. For example, in the well-developed area of soil classification, the basic taxonomy has supported derivation of various secondary classification systems, such as those describing land capability for agricultural purposes.

Primary classification systems for natural resource surveys should be hierarchical to permit inclusion or organization of information at different levels of generalization. Input data and user needs exist at levels ranging from global at one end of the spectrum to local and site-specific at the other. Hierarchical classifications provide a basis for designing surveys, mapping units, and analyzing at different scales and levels of detail. Many ecological classification systems are currently in use. An important and demanding task for a national biological survey would be to find ways of harmonizing or linking these to permit comparability of information.

A second aspect of an ecological inventory is the selection of methods, which can range from on-the-ground surveys to the use of aerial photography and satellite imagery. Recent developments in remote sensing, photointerpretation, and computer applications have revolutionized ecological inventories by dramatically improving the capability to collect reliable information over vast land areas and to prepare maps or digitized information at various scales. Yet many features still require extensive ground observations—for example, the presence or absence of various species.

Finally, and most difficult, is establishment of quantitative relationships between ecosystem types and various species of interest. For example, determining the habitat value of various ecosystem types for wildlife may require extensive analyses of the life requirements for food, cover, water, and reproductive habitat of individual species. Such information may be needed to assess the environmental impact of various proposed developments and management schemes.

Over five years ago, a review of federal agency wildlife habitat inventories and needs stated:

> "...the National Wetlands Inventory is systematically gathering information on the extent and distribution of various wetland types in accordance with a hierarchical classification system. The need now is for a systematic method of describing all values (for example, ecological, hydrologic, etc.) associated with each wetlands type. Considerable information is available on the wildlife associated with, or dependent on, some wetlands types and complexes of wetland types; less on others. What is required is a systematic means of structuring and cataloging such information, so that it can be usefully correlated with the information resulting from the Inventory on extent and distribution of wetlands at different levels in the hierarchical classification scheme." (Hirsch et al., 1979)

Despite intense interest in wetlands protection, that need is still unmet today.

An important niche that a national biological survey could fill is a means of systematically correlating the environmental requirements of various plant and animal species, in concert with ecological classification systems. If the systematics studies sponsored by the survey include information that help define species-habitat relationships, they will be of increased usefulness in environmental protection.

INFORMATION MANAGEMENT

Recognition that information management is a central and vital function of the proposed national biological survey is reflected by the prominent role of that topic on the agenda of this symposium. An effective information management program would help make existing biological information available and useful to environmental managers and could help develop greater comparability and consistency in future data collection efforts. Because this topic will be covered extensively elsewhere, I want to make only several points concerning information management here.

First, with respect to data base management, it is critical that such systems be designed with user needs in mind. For example, for some applications, data bases organized along taxonomic lines can serve environmental management needs. There has been considerable effort to develop statewide fish and wildlife species data bases that can address such topics as distribution of animals, their habitat requirements, and the response of animals and habitat to alternative land uses and management practices.

For other applications, systems organized along taxonomic lines would completely fail to meet management needs. For example, an existing biological information system was designed for use in water pollution control, but was organized principally along taxonomic lines. Data was entered and stored by species designation. The system therefore could provide information about the distribution of individual species of benthic invertebrates and plankton. However, for purposes of most water pollution analyses, information is needed about conditions by river reach. What is needed is a composite of biological information, water quality data, and data concerning waste discharges to provide an integral picture of conditions for each river reach. Thus the utility of the existing system is sharply limited, since the systems design failed to adequately consider user need.

Second, it is essential that the biological survey not only collect and store information but that it also synthesize such information to make it useful to the environmental manager. This may involve preparation of interpretive reports that characterize ecosystems or communities and their likely responses to stress, on the basis of the compilation of all available information from various sources. An example is the FWS community profile series (Jaap, 1984).

Information can also be made available through interactive computer systems and models that permit data manipulation to assess environmental impacts or the results of various management policies. Applied ecology has made significant strides in the past decade, with developments in remote sensing, photo interpretation, computer graphics, and simulation modeling. Increased availability of personal computers provides means of making these techniques accessible and useful to environmental managers as well as to researchers.

Environmental scientists are also beginning to explore applications of artificial intelligence, such as computerized expert systems that capture the information and problem solving processes of experts for use by less specialized personnel. If we cannot envision the survey going that far towards "high technology", at least initially, it could provide a useful service by simply maintaining registers of biological experts.

CONCLUSIONS AND RECOMMENDATIONS

1. To be most useful for application to environmental protection, a national biological survey should include information on population biology and ecology as well as taxonomy. Ecological information can be structured best within the framework of a hierarchical ecosystem classification scheme, which will strengthen our ability to apply research results and management experience to situations with similar properties. The survey should include inventory systems using remote imagery and photo-interpretation along with geobased information systems to store and manipulate the data. Taxonomic, life history, and species distribution studies could be fit within the framework of such a system, in effect combining a top-down bottom-up approach.
2. A wealth of relevant information is already being collected by universities, government agencies, and other sources, but existing information is often characterized by lack of comparability and accessability. An important and demanding first step in organizing a national biological survey should be to focus on means of making existing information available and on promoting consistency and comparability.
3. In addition, it is important that the survey analyze and interpret available information to provide reports relevant to the needs of environmental managers. Examples are reports showing shifts in the distribution of species, communities or ecosystems over time and ecological profiles that synthesize existing information on the ecology of selected ecosystems or communities.
4. Strong emphasis also should be placed on development and use of modern information systems technology and information networking to provide access to existing information sources through on-line systems, interactive models, registers of experts, and such emerging techniques as expert systems. Interactive information systems that can enable managers to address questions concerning the distribution of flora and fauna and their likely response to environmental stress are important products.
5. To remain viable and relevant, the national biological survey should incorporate strong provisions for interaction with the user community and for providing services and output meaningful to that community as well as to the scientific community. Further, to be successful and to grow, the fledgling survey must make useful information products available early on, even while it is laying the base for longer term accomplishments.
6. As compared with a more traditional taxonomic effort, a survey designed along the above lines would be very complex and expensive and would involve a number of formidable conceptual as well as managerial problems. It could not accomplish these goals overnight any more than the U.S. Geological Survey has established a framework for geological information in the U.S. overnight or than the U.S. Soil Conservation Service has done for soils. However, the planning framework for this broader effort could be established from the outset. The benefits would be much greater utility and a broader

base of support. This is clearly a key tradeoff that the organizers and sponsors of the national biological survey will have to make in determining the scope and objectives of this important endeavour.

LITERATURE CITED

Beanlands, G. E. & P. N. Duinker. 1985. *An ecological framework for environmental impact assessment in Canada.* Institute for Resource and Environment Studies, Dalhousie University and Federal Environmental Assessment Review Office, ISBN 0-7703-0460-5.

Hirsch, A., W. B. Krohn, D. L. Schweitzer, & C. H. Thomas. 1979. Trends and needs in federal inventories of wildlife habitat. *In: Transactions of the 44th North American Wildlife and Natural Resources Conference.* Wildlife Management Institute, Washington, DC.

Jaap, W. C. 1984. *The ecology of the South Florida coral reefs: a community profile.* U.S. Fish and Wildlife Service, Minerals Management Service, FWS/OBS-82/08, MMS 84-0038. U.S. Department of the Interior, Washington, DC.

Jones, P. W., J. D. Hardy, Jr., G. D. Johnson, R. A. Fritzche, F. D. Martin & G. E. Drewry. 1978. *Development of fishes of the Mid-Atlantic-Bight: an atlas of egg, larvae and juvenile states.* U.S. Fish and Wildlife Service, FWS/OBS 78/12. U.S. Government Printing Office, Washington, DC.

Judy, R. D., Jr., P. Seeley, T. M. Murray, S. C. Svirsky, M. Whitworth, & L. S. Ischinger. 1984. *1982 National fisheries survey.* U.S. Fish and Wildlife Service, U.S. Environmental Protection Agency, FWS/OBS-84/06. U.S. Government Printing Office, Washington, DC.

Karr, J. R. 1981. Assessment of biotic integrity using fish communities. *Fisheries* 6(6):21–27.

National Science Foundation. 1977. *Long-term ecological measurement. Report of a conference.* Woods Hole, Massachusetts, March 16–18, 1977.

Resh, V. H. & J. D. Unzicker. 1975. Water quality monitoring and aquatic organisms: the importance of species identification. *J. Water Pollut. Control Fed.* 47(1):9–19.

Tiner, R. W., Jr. 1984. *Wetlands of the United States: current status and recent trends.* U.S. Fish and Wildlife Service, U.S. Government Printing Office, Washington, DC.

U.S. Environmental Protection Agency. 1973. *Biota of freshwater ecosystems.* Water Pollution Control Series, 18050. U.S. Government Printing Office, Washington, DC.

Agricultural Research: the Importance of a National Biological Survey to Food Production

Waldemar Klassen
Agricultural Research Service

Abstract: For agriculture to become more efficient and more sustainable, we must be in a position to rationally manage not just crops and livestock, but also all beneficial and harmful organisms in the agroecosystem. A national biological survey would contribute to our ability to readily identify all agriculturally significant organisms (both before and after harvest) and to elucidate their roles and interactions.

Specifically, a national biological survey would be directly beneficial to agriculture by 1) providing information on germplasm resources that may be used to improve existing species of crops and livestock and to develop entirely new species of crops and livestock, 2) providing knowledge needed to a) enhance the activities of beneficial symbionts in essential processes such as nitrogen fixation, b) enhance the ability of mycorrhizae on or in roots of crops to facilitate the absorption of nutrients, and c) enhance the ability of microflora to breakdown crop and pesticide residues at appropriate rates, 3) facilitating the use in crop production of additional species of indigenous pollinating insects, 4) enhancing the actions of natural enemies of plant pathogens, nematodes, insects, and weeds in agricultural ecosystems, 5) identifying and coping with the tens of thousands of species of indigenous and immigrant pests in U.S. agriculture, and 6) coping with numerous pathogens and arthropods that reduce the value of harvested products and sometimes cause importing countries to deny entry of U.S. products.

Overall, a national biological survey would contribute to increased production efficiency, more effective and less detrimental use of fertilizers and pesticides, reduced losses caused by pests, reduced tillage, reduced soil erosion, more efficient use of water, improved air quality, better plant and animal health, more nutritious and wholesome food, and an improved balance of trade.

Keywords: Efficiency, Perpetual Sustainability, Ecological Impacts of Agriculture, Germplasm, Pollination, Nitrogen Fixation, Crop Improvement, Pest Management, Mycorrhizae.

The purpose of this paper is to discuss the importance of a proposed national biological survey in relation to food production in the U.S. Agriculture is the production of food, fiber, and other materials by means of the controlled use of plants and animals (Spedding et al., 1981). Agriculture is always concerned with two things, namely the efficient use of inputs and perpetual sustainability (de Wit et al., 1985). Thus, as a minimum, the net output of calories in agriculture must exceed the net input of calories supplied by muscle and machine power. We need crops and livestock that more efficiently convert nutrients into desired products. The ability of crops and livestock to be more efficient is affected by tens of thousands of species of beneficial and harmful organisms that share agricultural ecosystems with our favored economic species of crops and livestock. Thus, the rational management of the agroecosystem should be based on knowledge of *all* of its living components and on their beneficial and harmful roles and interactions.

Agriculture is neither practiced under asceptic conditions nor in total isolation from natural ecosystems; agricultural ecosystems impact on natural ecosystems and natural ecosystems impact on agricultural ecosystems. For example, some of the nitrogen that is added as fertilizer to agricultural fields eventually also fertilizes wilderness areas because microorganisms convert crop residues and animal manures into ammonia which precipitates in rain (Ryden and Garwood, 1984). Similarly, some pesticides may move far from the fields in which they are applied. Such movement of agricultural chemicals is wasteful, and much can be done to reduce it. Nevertheless, it is important to be able to accurately assess the extent to which agricultural uses of fertilizers and pesticides affect ecological processes in non-agricultural areas.

On the other hand, natural ecosystems export many pests to agricultural fields. Crops in the western U.S. are particularly vulnerable to invasions of pathogen-bearing arthropods from natural ecosystems. Examples include the pathogens of citrus stubborn disease and sugarbeet yellows, both of which are transmitted by leafhoppers that invade agricultural fields from the adjacent desert vegetation. Similarly, various species of wildlife serve as reservoirs for various viral, bacterial, protozoan, and helminthic diseases of livestock. Clearly, a national biological survey would provide a basis for understanding better the functioning of both natural ecosystems, and of agricultural ecosystems, and it would help in analyses of the impact of the two categories of ecosystems on each other. Moreover, a national biological survey would help in analyses of the impacts of industries and cities on both natural ecosystems and agroecosystems.

A national biological survey would be directly beneficial to agriculture by providing information on germplasm resources that may be used to improve existing species of crops and livestock and for developing entirely new species of crops and livestock. The survey could result in knowledge needed to a) enhance the activities of beneficial symbionts in essential processes such as nitrogen fixation, b) enhance the ability of mycorrhizae on or in roots of crops to facilitate the absorption of nutrients, and c) enhance the ability of microflora to break down crop and pesticide residues at appropriate rates.

The survey could result in the use in crop production of additional species of indigenous pollinating insects. This will become especially important if the Africanized bee and certain mite parasites of the honey bee become established in North America.

A national biological survey would be very helpful in enhancing the actions of natural enemies of plant pathogens, nematodes, insects, and weeds in agricultural ecosystems. The survey would help us to identify and cope with the tens of thousands of species of indigenous and immigrant pests in U.S. agriculture. The survey could help us to cope with numerous pathogens (Watson, 1971) and arthropods that reduce the value of harvested products and sometimes cause importing countries to deny entry of U.S. products. Overall, a national biological survey would contribute to increased production efficiency, more effective and less detrimental use of fertilizers and pesticides, reduced losses caused by pests, reduced tillage, reduced soil erosion, more efficient use of water, improved quality of air, better plant and animal health, and more nutritious and wholesome food.

I wish to expand somewhat on these points. With regard to crop and livestock germplasm, contemporary agriculture is still highly dependent on the decisions that were made by our ancestors during the Stone Age as to which species of plants and animals to domesticate for food and fiber production. In Europe, wheat, barley, millet, and flax were domesticated during the Neolithic Age, whereas oats and field beans were domesticated during the Bronze Age.

The Indians of the Americas introduced the early European settlers to maize, white potatoes, sweet potatoes, tobacco, peanuts, some varieties of squashes, field pumpkins, sunflower, Jerusalem artichokes, tomatoes, garden peppers, pineapples, watermelons, and various medicinal plants (Klages, 1949; Train et al., 1957). About one-third of U.S. agriculture is native American.

In recent centuries, we have profoundly improved the crops and livestock that were domesticated before the dawn of history. Yet in recent decades, only a few wild species have been adapted as significant food crops, an example being blueberries. Instead, most of the new plant species that have been domesticated in the U.S. in recent decades are floral and landscape plants.

Only 3 percent of the world's vascular plant species, or about 10,000, have been examined, at least superficially, as potential sources of vegetable oil, fiber, gums, antitumor agents, and high-energy sources (Agricultural Research Service, 1982a).

Many traits possessed by wild species would be useful in domestic plants. These traits include increased photosynthetic efficiency, tolerance to drought, frost, toxic soils, pests, etc. By means of embryo rescue via tissue culture, it is now possible to obtain useful hybrids from very wide crosses. Eventually, genetic engineering techniques will be used to transfer chromosomes and chromosomal genes between very divergent taxa. By means of protoplast fusion techniques, the genes that reside in plant mitochondria and in chloroplasts can be transferred between some taxa.

There is a need for many additional species of landscape plants, especially on the Great Plains and in our desert communities. Although many native U.S. trees and shrubs have been used for landscape planting since the earliest settlements, to this day many native species are little known or appreciated for their potential as landscape plants. As shown in Table 1, even those landscape plants now in cultivation have been relatively little exploited with respect to their germplasm potential. These plants include oak, maple, birch, ash, pine, elm, hickory, and others.

The adaptability of U.S. germplasm favors native U.S. plants for landscape use. This cannot be said for exotic species brought into the country, many of which

Table 1. Use of some native U.S. trees in the landscape. The number of U.S. species is derived from Elbert L. Little, Jr. *Checklist of United States Trees* (1979): The number of species in the trade (column 2) has been derived from *Sources of Plants and Related Supplies* (1979–1980 edition), published by the American Association of Nurserymen. (*) Genera with a few known cultivars.

	No. U.S. species	No. U.S. species in the trade
Abies (True fir)	9	5
**Acer* (Maple)	11	6
**Aesculus* (Buckeye)	6	1
Alnus (Alder)	8	1
**Betula* (Birch)	7	4
Carya (Hickory)	11	1
Catalpa (Catalpa)	2	1
**Celtis* (Hackberry)	5	2
Chamaecyparis (White cedar)	3	3
**Cornus* (Dogwood)	11	5
**Crataegus* (Hawthorn)	35	5
**Fraxinus* (Ash)	16	3
**Juniperus* (Juniper)	14	6
Liriodendron (Tulip tree)	1	1
**Magnolia*	8	3
**Malus* (Crabapple)	4	2
**Picea* (Spruce)	7	4
**Pinus* (Pine)	37	16
Platanus (Plane)	3	2
**Populus* (Poplar)	8	3
Prunus (Plum)	18	3
**Quercus* (Oak)	55	15
**Robinia* (Locust)	4	2
**Rhus* (Sumac)	11	2
Styrax (Storax)	3	0
**Ulmus* (Elm)	6	2

1. The above list includes some of the most important trees used in the U.S. for landscape purposes.
2. Genera with an asterisk (*) already have some known cultivars, nearly all produced in nurseries, not as a result of scientific breeding research.
3. Note the number of U.S. species as compared with the number of species in the nursery trade.
4. Scientific breeding research on any of the above named genera could generate valuable new selections and hybrids for the landscape, covering all parts of the U.S.
5. The above table was prepared by Dr. F. A. Meyer, U.S. National Arboretum, Washington, D.C.

have a narrow genetic base. The use of U.S. germplasm offers the possibility of sampling, at close range, the total complement of genetic variability of native species over their entire range. In a survey, this factor is essential for the establishment of a germplasm bank to assess genetic variability and test provenance.

In the U.S., we have made very great progress in developing the National Plant Germplasm System (United States Department of Agriculture, 1981). This system currently maintains well over 500,000 accessions of crops and their wild relatives in the form of seed and vegetatively propagated stocks.

The most serious weaknesses in the system include 1) inadequate number of trained collectors, 2) problems in identifying requested domestic material, and 3) the omission of domestic material in most collections.

Beneficial symbionts colonize or frequent agricultural fields. To become more efficient and sustainable, agricultural specialists must husband beneficial symbionts with the same care and effectiveness that is devoted to the crop. To husband these symbionts, we need to know their identities and roles.

The soil adjacent to plant roots is a zone of intense microbial and biochemical activity. This region, known as the rhizosphere, is the most complex and least understood part of the edaphic environment influencing plant vigor and health. It is the region where mineral uptake, water extraction, root pathogen invasion, root exudation of readily soluble materials, release of sloughed cells, chelation and release of minor elements, and localized changes in reaction and red-ox potential take place. It is in the rhizosphere that complex interactions occur between blue-green algae and bacterial diazotrophs including *Spirillum, Anabeana, Bacillus enterobacter, Azotobacter,* and others (Hardy et al., 1975).

Some beneficial fungi known as mycorrhizae live in very close association with the roots of plants. The mycelia of the mycorrhizae radiate into the surrounding soil and thereby greatly expand the effective surface of the root for increasing water and nutrient absorption. A bacterial flora is also observed in association with the mycorrhizae. Mycorrhizae may also protect crops from infection by pathogens such as *Phytophthora* (Baker and Cook, 1974).

The taxonomy of mycorrhizae has yet to be fully developed. Many endomycorrhizae cannot be cultured *in vitro.* Taxonomic, ecological, and physiological research is urgently needed as a basis for enhancing the beneficial activity of mycorrhizae in agricultural fields.

Microorganisms in the soil provide plants with nutrients by the cycling of minerals by decomposing plant and animal matter (Lynch, 1983). Agricultural scientists are investigating the effects of agricultural residue management on microbiological properties of soils and the interactions of these microbiological properties on the conservation and productivity of soils, on plant health and nutrition, and on the quality of crops. Such studies are particularly germane as more and more farmers are discarding the plow and adopting conservation tillage practices.

We rely on the microbial degradation of pesticides in the soil to guard our ground water supply from pesticide contamination. Scientists are attempting to identify and develop improved microbial forms for this purpose. On the other hand, in some soils, microbial populations have assembled that degrade pesticides so rapidly that harmful target species are not affected. We are in our infancy in terms of understanding and dealing with these "problem soils" (Kaufman and Edwards, 1983).

Pollination by bees and other insects is needed to assure seed production in many crops (Agricultural Research Service, 1976). The improvement of pollination would increase yields in the order of 10 percent (Agricultural Research Service, 1976). The introduced European honey bee is a good general pollinator but is less effective than some solitary bees on certain crops. Solitary bees are distributed among nine families, and perhaps many species could have an enhanced role in agriculture (Batra, 1984). In the northwestern U.S. and Canada, two species of solitary bees—the alfalfa leafcutter bee and the alkali bee—are now being managed intensively to aid in production of alfalfa seed. These examples suggest the magnitude of benefits that would accrue in other crops if we had a better knowledge of native bees.

Because of the anticipated invasion of North America by the Africanized honey bee and by various exotic mite parasites of the honey bee, it seems likely that interest in greater utilization of indigenous pollinators will mount substantially.

One of the most dramatic examples of the control of a pest by its natural enemies occurred in Los Angeles about 100 years ago. At that time, the cottonycushion scale had become established and was destroying the citrus industry. Its natural enemy, the vedalia beetle, was introduced from Australia. In short order, the vedalia beetle reduced the cottonycushion scale population to an inconsequential level and assured the future of the citrus industry in southern California. Since that time, there have been numerous successes in the practical biological control of insect pests by (1) importation of natural enemies from abroad, (2) conservation of indigenous natural enemies, and (3) augmentation of indigenous natural enemies by releases of mass-reared parasites and predators.

It is not uncommon to find 30,000 to 50,000 parasites and predators in an acre of cotton or soybeans. The ecology and relative importance of these beneficial insects is poorly understood. Nevertheless, Pimental et al. (1980) estimated that the inadvertent destruction of a portion of them by insecticides had increased the annual cost of insect control by almost $300 million in the U.S.

During the past decade, a number of exciting developments occurred in the biological control of plant diseases and weeds. Indigenous fungi have been discovered that strongly suppress soilborne disease agents including *Rhizoctonia, Fusarium, Pythium, Sclerotinia,* and *Verticillium* (Agricultural Research Service, 1984; Papavizas and Lewis, 1981).

Bacillus thuringiensis has been found to control *Cercospora* leafspot of peanuts, *Alternaria* leafspot of tobacco, and brown rot disease of peaches and other stone fruits. *Bacillus subtilis* gives reasonable control of bean rust on highly susceptible cultivars of dry bean and snap bean (Baker et al., 1985). Similarly, a bacteriophage has been shown to effectively control certain pathovars of *Xanthomonas campestris,* which causes bacterial leafspot on peaches and apricots (Randhawa and Civerolo, 1985).

Even certain viruses may be subject to an intracellular form of biological control. Kaper and Waterworth (1981) showed that a certain RNA satellite (CARNA 5) of cucumber mosaic virus can suppress replication of the virus. These results were applied in China by Tien and Chang (1983), who used protective inoculation with CARNA 5 to control tomato mosaic virus and cucumber mosaic virus in tomato and pepper, which resulted in yield increases of up to 60 percent.

Some weeds are effectively controlled by insects. This has been dramatically demonstrated by the importation, release, and establishment in the U.S. of the insect enemies of exotic rangeland and aquatic weeds. Well-known examples include the control of Klamath weed by *Chrysolina* beetles and control of alligator weed by several species of beetles. However, until recently the possibility of using biological agents against cropland weeds was not seriously considered because biological agents must be host-specific. A cultivated field normally is infested with at least 20–30 species of weeds, and the removal of one species results in its immediate replacement by the other weed species. However, the biological control of weed species has been shown to be feasible in rice and soybean fields by means of mixtures of fungal pathogens (Emge and Templeton, 1981).

Recent progress has shown the possibility of using species of the genera *Alternaria, Septoria, Colletotrichum, Fusarium, Protomyces,* and *Cephalosporium* to control weeds (Agricultural Research Service, 1984). Nevertheless, the surface has scarcely been scratched in identifying the indigenous organisms whose activites might be utilized in weed control.

The advent of the release of recombinant DNA into the environment, coupled with mounting concern over endangered species, is causing a thorough review of policies and procedures involved in the release of alien or exotic biological control organisms in the U.S. A national biological survey would provide much valuable information that is needed to develop Environmental Assessments and Environmental Impact Statements that are required for the release of genetically-engineered organisms. As shown in Table 2, about 16,000 species in the U.S. are known to be pests. These pests include insects, disease pathogens, weeds, nematodes, snails, slugs, birds, rodents, and other animals (Sutherland et al., 1984). Approximately 1,000 of these species cause severe losses each year (Agricultural Research Service, 1984).

U.S. agriculture has disrupted natural ecological systems on the same order of magnitude as occurs in the wake of major geological changes. We have replaced diverse vegetational assemblages (prairie, deciduous forests, deserts, etc.) with vast monocultures. Natural vegetation that has been left undisturbed by the plow is exploited at intensities unlike those posed by herbivores in the past (Whitcomb, personal communication). Simplification of ecosystems by the practice of monoculture in agriculture favors many pest species. In addition, many pest species have been introduced into new areas without their natural enemies. Therefore, pests are often more damaging in newly invaded areas than in areas where they originated. On the other hand, many species of plants and breeds of domestic animals have been introduced into new areas where they are attacked by a new set of species against which they have no innate defenses.

The continental U.S. is very vulnerable to the introduction of exotic species, many of which are pests. During the past 500 years, the Atlantic and Pacific Oceans have become progressively less effective as geographic barriers to invasion by exotic species. According to Sutherland et al. (1984), there are about 13,000 known species of alien pests awaiting transportation into North America. About 6,000 of these species would probably be significant in North America. A prior report by R. C. McGregor (1973) stated that there are 1,333 significant foreign pests of which 22 are animal pathogens, 551 are plant pathogens, and 760 are insects and other arthropods. A national biological survey would provide much of the data needed to distinguish between indigenous and emigrant pests.

MAGNITUDE OF A NATIONAL BIOLOGICAL SURVEY IN RELATION TO AGRICULTURAL NEEDS

The magnitude of the need for a survey in relation to agriculture can be judged from the following discussion:

A. *Prokaryotes*–Prokaryotes lack a nuclear membrane and organelles such as

Table 2. Numbers of species in major taxa that include pests.[1]

	Native pests in U.S.A.	Immigrant pests established in U.S.A.	Pests not known to occur in U.S.A.	Potential pests not known to occur in the U.S.A.	Total world pests	Pest and nonpest species[2]: Estimated total no. world spp.	Estimated no. described world spp.	Estimated total no. U.S.A. spp.	Total no. U.S.A. described spp.
Insects	5,000	865	600	6,000	10,000	10–30,000,000	751,000		87,100
Mites	350	150	300	400	1,200	1,000,000+	30–40,000	20–30,000	7,000
Molluscs	55	50	100	900	950	46,800	42,000	6,000	5,500
Animal disease microorganisms[3]	114	300	113	500	527	100,000–160,000[4]	24,000[4]	20–30,000[4]	6,000[4]
Plant pathogenic fungi	8,000	2,000	4,000	5,000	14,000	250,000	70,000	100,000	29,000
Plant parasitic nematodes	950	50	2,000	13,000	30,000	108,000[5]	18,200	35,000	6,000
Plant pathogenic microbes[6]	400	100	100	500	2,000	2,000	700	1,000	500
Vascular Plants	500	1,900	4,700	2,000	8,100	238,000	215,775	21,750	21,600
Vertebrates	120	86–111				86,300	85,900	6,069	5,219
B. Numbers of species in best-known major taxa, for comparison									
Butterflies	7+	1				17,500	16,500	763[6]	763[6]
Birds	7	3	30–100	30–100	30–100	9,000	8,600	1,106	1,106
Mammals	25	8				4,300	4,170	500	490
Reptiles	0	0				7,000	6,000	400	369
Amphibians	0	2				6,000	4,300	300	223
Fish	88	75–100				60,000	40,000	3,000	2,268

[1] Pest is defined as a species deleterious (causing economic loss or other negative impact, e.g., competitive exclusion of native species) to man, useful plants or domestic animals or likely to be deleterious. To qualify as a "pest," a species does not need to be serious enough to warrant control actions on a regular basis.
[2] For columns 6–9, the total number of pests and nonpests are included.
[3] Domestic animals (cattle, sheep, goats, horses, swine, poultry). Includes only Protozoa and helminths.
[4] Includes only helminths and protozoan parasites of mammals and birds.
[5] Includes plant parasites, free-living and predaceous forms in soil, parasites and associates of insects, and marine nematodes.
[6] United States and Canada.
Note: Sources of information are available from the compilers, L. Knutson, D. W. S. Sutherland, and R. L. Johnson.

mitochondria. They include bacteria, spiroplasmas and rickettsia-like bacteria. The comprehensive cataloging of prokaryotes of interest to agriculture would be a formidable task. Yet a comprehensive catalog would be of immense value in managing plant and animal diseases, ice-nucleation as a cause of frost injury, growth promotion in vascular plants, nitrogen fixation, decomposition of organic matter, etc.

B. *Fungi*–Only a small fraction of the species of fungi have been described. For example, there are roughly two and one-half times more species of rust fungi in our National Fungus Collections than have been reported in the literature on plant pathology. In recent studies on the biological control of plant pathogenic fungi, the most effective biocontrol fungi have been ones that are new to science. There is a vast amount of taxonomic and ecological work needed on microfungi, soilborne fungi, plant pathogenic fungi, mycorrhizal fungi, and endophytic fungi (Batra et al., 1978).

C. *Helminths*–Helminths include flukes, tapeworms, and roundworms. Only about 10 percent of the total known number of known taxa of nematodes have been described, i.e., 3,000 species are known and 30,000 species probably exist (Norton, 1978; Golden, 1968; Geographic Distribution Committee, 1983). A better knowledge of the nematode parasites of wildlife is needed to better understand the epidemiology of parasitic diseases of domestic ruminants (Al Sagur et al., 1982). Furthermore, the goal of developing truly phylogenetic classifications, with the predictive powers they entail, for the families of important livestock parasites cannot be achieved without including species parasitic in wildlife (American Association of Veterinary Pathologists, 1983; Lichtenfels et al., 1983; Lichtenfels and Pilitt, 1983).

D. *Arthropods*–Approximately 129,000 arthropod species have been described from North American (Agricultural Research Service, 1982b). Most species have been described only from a single stage, usually the adult, even though economic damage in most cases is caused by the unidentifiable immature stages. Thus Kosztarab (personal communication) estimated that for the insects in North America there is a need for about 900,000 new life stage descriptions, assuming that equal numbers of species remain undescribed as are described and that most species have four developmental stages and two sexes. In addition to insects, about 14,000 other terrestrial arthropods are known from North America, such as mites (including chiggers), ticks, spiders, and so forth. In addition, we need to bear in mind that field workers other than professional taxonomists are able to identify less than one fourth of the arthropod species.

E. *Vascular Plants*–About 22,000 vascular plants have been described from the U.S. There are several regional floras of substance, either complete or in preparation, and many state floras. Collectively, they provide tools for the identification of most U.S. plants, but many suffer because they are out-of-date and of limited geographic scope. Different taxonomic concepts have been applied by different workers.

Clearly, the knowledge of our national flora inadequately supports such ventures as describing germplasm diversity in relation to the development of new crops, providing a sound basis for identifying and preserving endangered species, supporting the biological control of weeds, and identifying and coping with poisonous plants on pasture and range.

RECOMMENDATIONS

1. Since the agricultural research community has both significant taxonomic resources and pressing needs for information and germplasm, persons involved in a national biological survey should maintain close communication and interaction with agricultural research scientists.
2. A national biological survey should provide for the application of survey data to elucidating the ecology of both managed and unmanaged ecosystems as well as the interactions between these two categories of ecosystems.
3. A national biological survey should give a relatively high priority to obtaining data needed to develop identification manuals of organisms that are significant or potentially significant in agroecosystems.
4. A national biological survey should provide data as well as preserved and living material relevant to agricultural interests in germplasm, beneficial symbioses (especially in the rhizosphere), pollination, harmful organisms, and natural enemies of harmful organisms.

In conclusion, for agriculture to become more efficient (as measured by calories of output minus calories of input) and for agriculture to be more sustainable, we must be in a position to rationally manage not just crops and livestock, but also all beneficial and harmful organisms in the agroecosystem.

To the extent that a national biological survey would contribute to our ability to readily identify all organisms that are significant to agriculture (both before and after harvest) and to elucidate their roles and interactions, a national biological survey would be a great benefit in helping the agricultural industry to provide the world's best food bargain to the American people and in making our products more competitive in the world market.

ACKNOWLEDGEMENTS

I am deeply grateful for helpful suggestions from Drs. E. L. Civerolo, R. E. Davis, R. A. Humber, L. Knutson, F. A. Meyer, W. L. Murphy, G. C. Papavizas, R. E. Perdue, A. Y. Rossman, J. R. Lichtenfels, and R. T. Whitcomb; and the persons who contributed information for Table 2, E. W. Baker, W. Dowler, A. M. Golden, C. R. Gunn, R. L. Johnson, R. J. Lichtenfels, W. L. Murphy, A. Rossman, and D. W. S. Sutherland.

LITERATURE CITED

Agricultural Research Service. 1976. *Crop pollination and honey production*, NRP No. 20180. 55 p.

Agricultural Research Service. 1982a. *Determine plant sources of useful raw materials and evaluate their crop and production potential in terms of U.S. agricultural and industrial needs.* Approach Element 3.34. Unpublished document.

Agricultural Research Service. 1982b. *Taxonomic research and services to support action programs.* Approach Element 3.34.1. Unpublished document.

Agricultural Research Service. 1984. *Research planning conference on biological control.* U.S. Government Printing Office. 473 p.

Al Sagur, J. Armour, K. Bairden, et al. 1982. Field study on the epidemiology and pathogenicity of different isolates of bovine *Ostertagia* spp. *Res. Vet. Sci.* 33: 313.

American Association of Veterinary Parasitologists. 1983. Research needs and priorities for ruminant internal parasites in the United States. *Am. J. Vet. Res.* 44: 1-836.

Baker, C. J., N. Mock & J. R. Stavely. 1985. Biocontrol of bean rust by *Bacillus subtilis* under field conditions. *Plant Dis.* (in press).

Baker, K. F. & R. J. Cook. 1974. *Biological control of plant pathogens.* W. H. Freeman and Co., San Francisco. 433 p.

Batra, S. W. T. 1984. Solitary bees. *Sci. Am.* 250 (2): 120–127.

Batra, L. R., D. R. Whitehead, E. E. Terrell, A. M. Golden, & J. R. Lichtenfels. 1978. *Overview of predictiveness of agricultural biosystematics.* Beltsville Symposia in Agricultural Research 2. Biosystematics in Agriculture. Allanheld, Osmun and Co., Montclair, New Jersey.

de Wit, C. T., H. Huisman & R. Rabbinge. 1985. *Agriculture and its environment: are there other ways?* In preparation.

Emge, R. G. & G. E. Templeton. 1981. Biological control of weeds with plant pathogens. *In*: G. C. Papavizas (ed.) *Biological control in crop production.* Beltsville Symposium in Agricultural Research. Allanheld, Osmund Publishers, Granada.

Geographical Distribution Committee. 1984. *Distribution of plant-parasitic nematode species in North America.* Published by the Society of Nematologists. 205 p.

Golden, A. M. 1968. Nematology: Our society and science (Presidential Address). *Nematology News Letter.* 14: 2–8.

Hardy, R. W. F., P. Filner & R. H. Hageman. 1975. Nitrogen input. *In: Crop productivity research imperatives.* Mich. Agric. Exp. Station and C. F. Kettering Foundation. Yellow Spring, Ohio. 399 p.

Kaper, J. M. & H. E. Waterworth. 1977. Cucumber mosaic virus-associated RNAs: causal agent for tomato necrosis. *Science* 196: 429.

Kaufman, D. D. & D. F. Edwards. 1983. Pesticide/microbe interaction effects on persistence of pesticides in soil. *In*: Proc. pesticide chemistry: human welfare and the environment. Vol. 4 of J. Miyamoto and P. C. Kearney, eds. *Pesticide residues and formulation chemistry.* Pergamon Press, Oxford.

Klages, K. H. W. 1949. *Ecological crop geography.* The Macmillan Company, New York. 615 p.

Lichtenfels, J. R., K. D. Murrell, & P. A. Pilitt. 1983. Comparison of three subspecies of *Trichinella spiralis* by scanning electron microscopy. *J. Parasitol.* 69: 1131–1140.

Lichtenfels, J. R. & P. A. Pilitt. 1983. Cuticular ridge patterns of *Nematodirella* (Nematoda: Trichostrongyloidea) of North American ruminants, with a key to species. *Syst. Parasitol.* 5: 271.

Lynch, J. M. 1983. *Soil Biotechnology.* Blackwell Scientific Publications, Oxford. 191 p.

McGregor, R. C. 1973. *The emigrant pests.* Animal Plant Health Inspection Service. 167 p.

Norton, D. C. 1978. *Ecology of plant-parasitic nematodes.* John Wiley & Sons, Inc., New York. 268 p.

Papavizas, G. C. & J. A. Lewis. 1981. Introduction and augmentation of microbial antagonists for the control of soilborne plant pathogens. *In*: G. C. Papavizas (ed.) *Biological control in crop production.* Beltsville Symposium in Agricultural Research. Volume 5. Allanheld, Osmund Publishers, Granada.

Pimentel, D., D. Andow, D. Gallahan, I. Schreiner, T. E. Thompson, R. Dyson-Hudson, S. N. Jacobson, M. A. Irish, S. F. Kroop, A. M. Moss, M. D. Shepard & B. C. Vizant. 1980. Pesticides: environmental and social costs. *In*: D. Pimentel & J. H. Perkins (eds.) *Pest control: cultural and environmental aspects.* AAAS Selected Symposium Westview Press, Boulder, Colorado. 243 p.

Randhawa, P. S. & E. L. Civerolo. 1986. Biocontrol of *Prunus* bacterial spot disease by purified high titre pruniphage. *Plant Disease* (in press).

Ryden, J. & E. Garwood. 1984. Evaluating the nitrogen balance of grassland. *In*: J. Hardcastle (editor). *Grassland Research Today.* AFRC, London, United Kingdom. 28 p.

Spedding, C. R. W., J. M. Walsingham & A. M. Hoxey. 1981. *Biological efficiency in agriculture.* Academic Press, London. 383 p.

Sutherland, D. W. S., L. V. Knutson, J. R. Dogger & R. Johnson. 1984. *The concept of a National Agricultural Pest Register.* Unpublished document.

Tien, P. & K. H. Chang. 1983. Control of two seed-borne virus diseases in China by the use of protective inoculation. *Seed. Sci. Technol.* 11: 1–4.

Train, P., J. R. Henrichs & W. A. Archer. 1957. *Medicinal uses of plants.* Quarterman Publication, Inc., Lawrence, Massachusetts. 139 p.

United States Department of Agriculture. 1981. *The National Plant Germplasm System: current status (1980). Strengths and weaknesses, long range plan (1983–1997).* Washington, DC. 166 p.

Watson, A. J. 1971. *Foreign bacterial and fungus diseases of food, forage and fiber crops: an annotated list.* U.S. Dept. Agric. Handbook 418. 111 p.

Plant Protection and a National Biological Survey

Ronald L. Johnson
Animal and Plant Health Inspection Service

Abstract: The Animal and Plant Health Inspection Service (APHIS) is engaged in various activities to protect U.S. agriculture from plant pests, especially those not known to occur in the United States. The first line of defense used by APHIS is to impose quarantines and intercept pests at ports-of-entry. In 1983, 40,689 pests and pathogens were intercepted. Since all quarantines are not 100 percent effective, a Cooperative National Plant Pest Survey and Detection Program (CNPPSDP) has been implemented as a second line of defense. The program coordinates existing surveys for insects, weeds pathogens, nematodes, etc., to collect, store on a central computer, process, and retrieve plant pest information. The program addresses plant pest information needs for endemic and exotic pests and for export certification. Emphasis is on a multiagency and multidisciplinary coordinated biological survey and detection effort.
Keywords: Survey, Detection, Quarantine, APHIS, Computer, Exotic, Export Certification, Endemic, Multiagency, Multidisciplinary.

INTRODUCTION

The economy of the United States, as well as that of the world, is extremely vulnerable to widespread adverse weather conditions or pest infestations. We are still incapable of controlling the weather, but modern research and defensive tactics have allowed us to limit pest damage in some instances. In an attempt to meet the world's food and fiber needs, man has enhanced his vulnerability to pests by changing nature's balance and creating extremely unstable monocultures. The balance has been further upset by mass transportation, which enables the rapid relocation of biotic agents.

The complexity of the problem can be appreciated when one considers that there are hundreds of thousands of described and undescribed species of insects, weeds, nematodes, pathogens, and other organisms and that thousands of different agricultural products enter dozens of U.S. ports each day. Add to these numbers the millions of travelers who enter these ports each year, and it is obvious that a very real risk to U.S. agriculture exists. Plant protection agencies in the U.S. Department of Agriculture (USDA) and the states employ methods to intercept and eliminate plant pests. These agencies must also be capable of properly identifying these pests and carrying out the quarantine actions dictated by the regulations of their country (Fowler, 1984).

In response to problems involved with the movement of plant pests, Plant Protection and Quarantine (PPQ), APHIS, USDA, was established. The primary function of PPQ is to protect the U.S. from the destructive activities of exotic and certain endemic plant pests. PPQ's initial defensive tactic is to try and stop the entry (movement) of exotic biotic agents into the U.S. Any quarantine or regulatory measures used to achieve this goal are meant to promote and protect agriculture, not to restrict trade.

In this presentation I will briefly describe the quarantine, regulatory, and control measures used by PPQ to reduce the movement of unwanted biotic agents and then summarize the goals and progress of the Cooperative National Plant Pest Survey and Detection Program (CNPPSDP).

OPERATIONS OF PORT OF ENTRY

As a first line of defense against the entry of exotic pests into the U.S., PPQ officers are stationed at ports of entry to inspect for and restrict the entry of unwanted species. PPQ employees at these ports apply regulatory and quarantine principles to imported agricultural products that enter the United States by land, sea, air, and even by mail. All agricultural products entering the U.S. from foreign areas and offshore locations are subject to inspection at the first U.S. port of entry. The purpose of this system is to determine the presence or absence of plant pests and prohibited agricultural materials (Fowler, 1984).

PPQ officers are trained in how and where to inspect for biotic agents. They are provided with lists of undesirable exotic plant pests, recognized hosts, and countries of occurrence. When officers are confronted with a pest or host, they have several means of action at their disposal. The most common actions are to inspect and release, to require that the infested items be treated, or to refuse entry of the item. For example, if a traveler attempted to import plant material prohibited by regulations because of potential harboring of latent viruses or other undetectable pests, these items would be seized and destroyed or refused entry. However, if the items were from a country in which no latent pests were known from that product and no pests were observed during inspection, the plant material would be inspected and released.

Some plants must be maintained through a post-entry quarantine growing period (commonly 2 years), during which time they are observed for any previously undetected exotic agents (for example, systemic plant diseases). Importers can apply through PPQ for additional information on permits and conditions of entry.

PPQ-approved treatment of infested goods or goods suspected of being infested is another option. A common treatment is fumigation by methyl bromide. This method is routinely used with infested cut flowers imported in commercial quantities. A less drastic measure is the safeguarding of potentially infested items transiting the U.S. to another country. PPQ applies safeguards to assure that the items pose no pest risks while moving through the U.S.

Exotic biotic agents do not respect national borders. This is evidenced by the fact that our PPQ officers intercepted 40,689 significant pests and pathogens in fiscal year 1983. In most instances, man or his belongings are the carrier. With numerous opportunities for pest entry—whether it be a result of modern transportation, the pests' own locomotion, or nature's help (i.e. wind, water, or other animals)—PPQ's first line of defense cannot always exclude exotic biotic agents.

PEST RISK ANALYSIS

Pest risk analysis is the process used to determine the conditions under which the importation of plants and plant products may be allowed, if at all. The sequence of events that leads to the development of an entry status determination for a given plant or plant product is:

1. A U.S. importer requests authorization to import a specific agricultural product. The application specifies the port in the U.S. through which the commodity will be imported, as well as the country of origin.
2. If pest risk has not been previously analyzed on that commodity from the country of origin, a formal analysis must be performed. The pest risk analysis will identify those ecomonic pests that may infest that commodity from that origin.
3. The decision to authorize or deny entry of the agricultural commodity will be based on the results of that analysis. If the permit is issued, the conditions under which the commodity may be imported are stated on the permit.

At times permit issuances are delayed because of misinformation in biosystematic literature, by synonymy, or by taxonomic name changes. Delays may also be experienced because of unconfirmed pest identification or errors in reported geographic distribution or host identification. The biosystematic needs of this program are rapid literature retrieval systems, current synonymy lists, rapid pest and host identification, and expanded host distribution information. The lack of information of this type in otherwise thorough literature searches may result in avoidable permit restrictions or denial of import authorization (Fowler, 1984).

SURVEY

PPQ's second line of defense is survey for the early detection of exotic insects, weeds, pathogens, and other pests. The need for early detection is reflected in the fact that U.S. crop losses caused by weeds alone were valued at $7.5 billion in 1979 (Chandler et al., 1984) with current crop yield losses averaging about 20 percent (Lackey, personal communication, 1985). Of the approximately 200 major weeds that infest cultivated crops in the U.S., 108 are of foreign origin (Shaw, 1968). Under the Federal Noxious Weed Act of 1974, an additional 93 weeds are listed as serious pests and are prohibited entry into the U.S. The results of survey and detection efforts may lead to programs that involve the eradication, suppression, or containment of a significant biotic agent should it become established in the U.S.

It is important that national pest surveys and identification support services be maintained, since no quarantine program will be 100 percent effective. While some of these quarantine and survey programs monitor established pests, many determine the presence of exotic or newly introduced pests. In programs aimed at exotics, the identification needs may be concerned with one specific insect, as in the case of *Ceratitis capitata* (Mediterranean fruit fly), or as in the case of citrus, the focus may be directed toward a wide variety of pest species. The survey and monitoring efforts of APHIS are involved with programs of both types. In these programs, information on plant pests is gathered by use of diverse trapping and monitoring techniques. Experienced taxonomists from cooperating state and federal agencies collect, identify, and record the organisms obtained from these field operations. The information is then processed and stored on automated data systems for use by scientists and agricultural communities worldwide. Precise taxonomic determination is required to define any actions that will be taken as a result of these programs (Fowler, 1984). I will address pest surveys further when I describe the CNPPSDP.

CONTROL

Eradication is usually the ideal but often unachievable goal of pest control. The methods used to achieve the goal are chemical and biological attacks with carefully regulated movement of the infested material. When eradication is unsuccessful, suppression or containment strategies are attempted next.

Suppression is the strategy of keeping the pest population below the economic threshold. This depends mainly on the use of chemical and biological techniques. Pennsylvania combats the gypsy moth with aerial spraying and the release of USDA-approved natural predators. On a smaller scale, individual farmers nationwide use suppression strategies when spraying their crops.

Containment is the strategy of stopping the expansion of an infested area. It is dependent on tight restrictions on the movement of materials out of the infested zone. For example, in North and South Carolina, all farm machinery leaving the witchweed quarantine area must be soil-free.

With our limited knowledge and abilities, neither suppression nor containment are realistic long-term strategies for dealing with exotic pests. However, they do provide much needed time for further research into more efficient chemical and biological control methods. They also provide a "catch up" time for natural parasites and predators to hold the pest in check.

COOPERATIVE NATIONAL PLANT PEST SURVEY AND DETECTION PROGRAM

The preceding information very briefly summarizes some preventive measures used by PPQ. However, because of limited funding, unavailability of pesticides, unsuitable biological agents, and restrictive state and federal laws and regulations, PPQ's results often fall short of our goals. The CNPPSDP was initiated to strengthen PPQ's ability to detect exotic agents at an early stage. Machiavelli said, "Before you can effectively attack the enemy, you must get to know his strengths and weaknesses." We are learning the strengths and weaknesses of plant pest survey and detection by gathering information on the harmful biotic agents in this country through the cooperative efforts of the many agencies, organizations, and individuals involved in survey and detection activities. Through the CNPPSDP we are attempting to determine precisely where these biotic agents are established, where they are moving to, where they have been found recently, and where they can be expected to be found in the future.

The Program was created in 1981 as a pilot program to demonstrate the feasibility of collecting plant pest and pathogen information at the state and local level, transmitting it via telecommunications to a central computer, storing the data, and providing a means for retrieving it on a timely basis. It is the intention of PPQ to use this Program to help us meet several of our legally mandated responsibilities, which include the protection of American agriculture from foreign plant pests introduced and established in the U.S. and the facilitation of movement of U.S. agricultural products in international commerce.

In PPQ our immediate goals are to:

1. Detect new and exotic plant pests of importance to American agriculture early enough to initiate an effective action program;
2. Compile information on important endemic plant pests in support of export certification; and
3. Provide timely information on the distribution and population levels of PPQ program pests.

The Program will assist PPQ-APHIS in meeting these responsibilities by providing a means of early detection, documentation, and rapid dissemination of information on exotic plant pests in the U.S. and on pests of concern for export certification. Information on the latter category will enable our officers to issue phytosanitary certificates on the basis of reliable and easily accessible data on certain endemic plant pests. This will provide better assurance that our agricultural products meet the entry requirements of other nations and will help improve our balance of trade.

There is yet another aspect of the Program, that of dealing with the other endemic plant pests and pathogens. Since the Program is cooperative, it is important that it provide benefits to all participants. The universities, especially their extension services and research stations, have a vital need for current data on the status of endemic pests. The Program provides a means for collecting data on these pests as well as those of concern to PPQ. This information will be used in monitoring endemic pest situations, such as first of season occurrence, distribution, economic threshold levels, trends in the movement and dispersal of various pests, and assessing crop losses.

We feel that a number of features of the Program are keys to its success. PPQ has negotiated a cooperative agreement with each of the 50 states. One of the requirements of the cooperative agreement provides that the state must establish a survey committee with representation from the state land grant university, state department of agriculture, local PPQ representatives, and others as appropriate. The latter may include representatives from other federal agencies, natural resource agencies such as forestry or conservation, state highway administrators, chemical company representatives, grower or commodity groups, private agricultural consultants, researchers, or others. We also require that the survey committee be multidisciplinary in that it represent the needs of entomology, plant pathology, weed science, nematology, and others. It is the responsibility of the state survey committee to determine the survey needs within that state and develop an annual workplan to meet those needs. We believe that through the cooperative efforts of the agencies, organizations, and individuals representing the different disciplines we will have the greatest opportunity to obtain rapid detection of exotic pests and pathogens and expand our data base on pests of concern for export certification. At the same time, we can maintain the type and quality of data necessary for those who need this information on endemic pests.

The state survey committee has the responsibility of developing a plan of work to address the survey needs to be carried out under the program. The various participants carry out their portion of the plan and submit their data to the state survey coordinator. The coordinator assures that the data is properly formatted and contains the required pieces of information. The data is then entered into the system. Emphasis is placed on entering only high quality data that is of value to neighboring states or for regional or national use.

All cooperators are able to access the data base to retrieve data or reports. Specific report formats are available for new pest records, first of season occurrence, trap data, pests of a single crop/host, crops/hosts for a single pest, diagnostic reports pest distribution, and pest and crop development over time.

When the new data base management system (DBMS) that is under development is implemented, ad hoc query will be available as a routine retrieval feature.

Another key element of the Program involves standardization of pest and pathogen data collection, storage, and processing techniques. The uniform storage and processing of data, at least at the national level, are accommodated in that all users of the system must input data according to a designated format using the data elements required by the system.

Standardization of survey methodology is another matter. We in PPQ do not feel we can or should dictate to the states the methods to be used for survey of a given pest or crop. We recognize, however, as do cooperators, the importance of being able to relate data from one source to data on the same crop or pest from another source. We are, therefore, encouraging the states in every way possible to work cooperatively to identify the survey methods for a given crop and pest situation. Through this process, we are urging adoption of the preferred method, thus establishing a mutually acceptable form of standardization and improving the quality of the collected data. The CNPPSDP requires that survey methods be reported along with other pest data. The national program includes a survey methods subfile called "Pest Survey Methods Information Database" (PeSMID) that will enhance standardization efforts by listing descriptions of the most commonly used survey methods and those contained in the literature. Having access to the survey methods used by others and mutually agreeing to use preferred survey methods will lead to expanded usability of the data.

The philosophy of this Program is to consolidate, as much as possible, survey efforts at the local, state, and national level rather than initiate a totally new national survey and detection program. The funding that we provide to the states is intended to enhance rather than to replace or duplicate ongoing survey and detection activities. It is our desire to collect data in the national system that is of importance regionally or nationally and to enhance it where necessary. We hope to take advantage of the several million staff hours of time devoted to survey and detection annually in the U.S. by having surveyors and scouts recognize the needs and interests of all cooperators and agree to collect specimens of unknown species and submit them to taxonomists for identification. This will place an additional burden on taxonomic resources but will certainly enhance our chances of finding exotic species or new infestations of pests not widely distributed.

To make maximum use of the data in the system, it is important that we develop a DBMS that will provide for maximum use of the data. PPQ is in the process of developing a DBMS for the Program. We have formed an Automated Data Processing Committee to provide technical guidance in this effort, and we have received input from all states to assure that the system meets as many of the needs as possible. We have completed the requirements analysis phase of the development process and are now designing the system. We expect to be on line with the DBMS in January 1986. At that time we will relocate the system from the Fort Collins Computer Center to a private time share facility—Planning Research Corporation—to merge with the National Pesticide Information Retrieval System. This will allow us to use the ADABAS software licensed for use by the NPIRS program, allow public access to the system, simplify access to the mutual users of both systems, and provide other intangible benefits to users.

Another key component for the national system involves the ability to use realtime weather data in connection with the currently available plant pest information. We are investigating the possibility of using existing weather data and developing, where necessary, the additional agricultural weather data necessary for these efforts.

The system will contain historic data of two types. First, past years' data will be stored on tape and will be accessible through the data base administrator. Second, a historic record of pest distribution to the county level will be maintained as a separate file. It will take some time to get that file up to date, but eventually new data will be checked against that file and it will be updated automatically.

One additional point I would like to make is that the Interamerican Institute for Cooperation in Agriculture (IICA) has requested help from PPQ in designing a plant and animal pest and disease reporting system for their 29 member countries in North, Central, and South America. I have consulted with IICA on that matter and expect someday to see a hemisphere-wide pest reporting system similar to our CNPPSDP. There is also other interest worldwide in our system as expressed by FAO in Rome and by several other countries.

There are weaknesses in the system, however, and several of them are of interest to this group. The taxonomic aspects of the program are a problem in several ways. First of all, the identification capabilities are very limited in many states, at least for certain families and genera of organisms. The program will certainly impact adversely on the taxonomists who must identify the additional specimens that are submitted as a result of the coordinated survey. Additionally, as we request or conduct special surveys for certain exotic species, such as a new pheromone trapping project we have initiated, we find that the number of taxonomists able to identify those species is very limited. Identification keys and aids are needed along with reference specimens. Frequently these are difficult to obtain or are unavailable.

A second problem involves coding pests and hosts. As you can appreciate, dealing with a single discipline can sometimes be difficult. With this program we are involved with all of the disciplines, and it is no small problem to confirm names of pests and hosts and assign a meaningful code. We are converting our data to the coding system used by the Environmental Protection Agency in an effort to standardize coding. That conversion throws an additional piece into an already complex puzzle.

Third, it is always best to know the biota that exists at a given point in time as a reference point from which to conduct surveys. As we know from our program, much data exists, but it is scattered, hard to find, and difficult to retrieve. In addition to helping to find new pests that are not now known to occur here, as seems likely, a national biological survey would establish that base point from which to conduct survey programs. We see a national biological survey as complementing the CNPPSDP and vice versa.

FUTURE METHODS

The resolution of taxonomic questions mentioned earlier will require studies that go beyond classical morphological consideration. It is apparent that techniques such as electrophoresis, cuticular hydrocarbon analysis, radioimmunoassay, monoclonal antibodies, venom analysis, and scanning electron microscopy must be used to solve some biosystematic questions (Fowler, 1984). The specific methods to be used in each of these techniques need to be further refined and described.

CONCLUSION

We feel the CNPPSDP will mature in several years to the point that it will provide the federal government, state cooperators, farmers, private industry, and others with the ability to obtain real-time plant pest information on a current basis and will enable them to meet their plant pest information needs and act accordingly.

RECOMMENDATIONS

1. Initiate a program to coordinate taxonomic resources and activities similar to the way survey and detection activites are being coordinated under the CNPPSDP (pull the pieces together).
2. Identify a base level of the biota of the U.S. such as would be accomplished through a national biological survey.
3. Encourage the cooperation of federal agencies to pool resources to find a pest survey system and identify mutual objectives.

LITERATURE CITED

Chandler, J. M., A. S. Hamill & A. G. Thomas. 1984. *Crop losses due to weeds in Canada and the United States.* Weed Science Society of America, May 1984.

Fowler, J. L. 1984. *Biosystematic needs in regulatory quarantine action programs.* Paper presented at the XVII International Congress of Entomology, August 1984, Hamburg, Germany.

Johnson, R. L. 1984. *The movement and dispersal of biotic agents and the role of quarantine and regulation measures.* Paper presented at the International Conference on the Movement and Dispersal of Biotic Agents. Baton Rouge, Louisiana, October 1984.

Lackey, J. 1985. Personal communication. PPQ-APHIS-USDA, Hyattsville, Maryland.

PPQ-APHIS-USDA. 1983. *List of intercepted plant pests, Fiscal Year 1983.*

Shaw, W. 1968. *The status of preventive weed control. Report of the Interagency Ad Hoc Committee on Preventive Weed Control.* USDA/USDI, June 3, 1984.

SECTION III.

BIOLOGICAL SURVEY INFORMATION

Prefatory Comments

Stanwyn G. Shetler
National Museum of Natural History

We are hearing about biological surveys from every angle in this symposium, with considerable but useful overlap. I merely want to preface this afternoon's session with a few thoughts from my own perspective.

What is "biological survey information"? More fundamentally, what is "information"? I will pass up the challenge to give a theoretical or philosophical answer and offer a simple, practical definition. "Information" is raw data that have been packaged in a meaningful way to serve some useful purpose—the end-product of organizing, integrating, synthesizing, and presenting the data for particular uses and users. The time-honored form of packaging is publication, but, living in a technological age as we do, we know that there now are other ways to package and present data. "Biological survey information," in the simplest terms, therefore, is information derived from biological data that have been collected through surveys.

The great repositories of biological survey data, especially collection-based data, are the world's herbaria and museums and the publications and other forms of information dissemination that emanate from these institutions. In a real sense, therefore, every herbarium or museum, large or small, is a biological survey. Obviously, the larger the museum the more comprehensive it is as a biological survey. A museum is a biological survey because it builds and houses biological collections, publishes useful works, and provides information to various publics through many forms of outreach. How well it transforms mountains of data into useful information is the measure of its value as a biological survey.

Why, then, with all the museums in North America that have been collecting data and disseminating information for many years—some dating to the early days of our country—do we now feel the need to create a "national" biological survey? Obviously, we have not done a very good job of transforming our data into clearly useful products, especially of a broad, integrated type. For too long we have operated as separate, individual and institutional fiefdoms, and our individual products have not amounted to anything approaching a uniform taxonomic and geographic coverage of the North American continent. A national biological survey, properly established, will provide us with the integrative concept and force to realize a truly national network with truly national objectives and, consequently, national support.

Where do we start? What should be the initial goals? The greatest risk that we face is to try to run before we can walk—to place so many initial information demands on the system that we overload it hopelessly before we ever start. It is clear already that we have enormous expectations. It is a risk of the political process of gathering support. I submit that we must maintain a sharp and relatively narrow data focus *at first* until some important initial information objectives are achieved or are well along toward being achieved. In my view, the single greatest need at this time—and the driving force behind the national biological survey movement—is the need for a continental fauna and a continental flora, in the form both of hardcopy publications and databases. I will not stick my neck out farther at this stage and try to define the data elements of faunas and floras but simply will emphasize the great power of a modest biological database to answer important questions when it includes a consistent data set across all taxa and areas. The ability to permute even a handful of variables for the entire fauna or flora could be enormously useful. Specifically, having a data set that would enable us merely to identify and name consistently all the plants and animals of North America would be of incalculable value, especially if this database were computerized so that one could use the power of the computer to identify an organism, starting from any data point.

Faunas and floras provide a rationale for synthesizing biological data over a wide taxonomic scope. They represent a comprehensive, if relatively shallow, horizontal dimension of data about many plants and animals, whereas detailed systematic and other biological studies provide the in-depth, vertical dimension of data that always proceeds slowly and opportunistically. In other words, faunas and floras give synthetic geographic scope; systematic and ecological studies give synthetic biological depth. Furthermore, the faunas and floras also provide the reference system—the framework of names— by which to catalog all species-related data. Producing this framework obviously is a first step in creating a biological data bank.

Finally, there is the big question of how a national biological survey should be organized to produce the information desired. I see several important steps. *First*, I think we need a new national legislative act, an organic act analogous to the Endangered Species Act, that defines a national biological survey and establishes a national mandate. Without this, a national biological survey will never have the public visibility and support it needs, and the federal agencies that might take part lack anything on which to hang requests to participate without jeopardizing their existing programs. *Second*, one or more agencies must be designated to seek appropriations to support a national center and the cooperating network of federal, state, and private agencies, institutions, and organizations, *and* for contracting with specialists to get the job done. I strongly favor contracts over grants, because I think a national biological survey will never reach its goals without some teeth to specify and control the production process. *Third*, I see a need for a thorough definition and planning phase during which, among other things, the status of our knowledge about the North American fauna and flora would be assessed and a report issued. Only after this planning phase would full-scale survey work begin.

Let me say that there is a considerable and growing interest in a national biological survey at the Smithsonian Institution and that the National Museum of Natural History, which I represent, has made such a survey one of its budget priorities for Fiscal Year 1987. In an interview published in *Science* magazine (June 28, 1985, p. 1512–1513), Smithsonian Secretary Robert McCormick Adams expressed interest in a national biological survey on behalf of the Institution.

As a concluding footnote, I remind you of the Flora North America (FNA) Program of the late 1960s and early 1970s with which I was intimately associated. In fact, for me, there is a strong sense of déjà vu in our deliberations at this symposium. For FNA, we had assembled all essential elements except for the *long-term* funding, and for a brief time we thought that we had this element in hand. Because we were so close, it was a great pity that we could not continue, but I think that FNA has left us with an important legacy. Of the many valuable lessons we learned, two are key in the present context: 1) that a project of the scope and permanence of the proposed national biological survey must have *long-term support assured* at the outset and, 2) that there is a genuine desire in the botanical scientific community at large for a focused, unified effort with national leadership to study the flora of the North American continent. I am certain of a similar groundswell of support from the biological community for the broader biological survey. FNA triggered rising expectations that have resulted in a sense of inevitability that will not die. People want it to happen and expect it to happen sooner or later. The move for a national biological survey can build on, and profit from, this tide of expectation.

Finally, let me urge that the geographic limits of the survey not be fixed at the U.S. boundaries. Plants and animals do not respect national boundaries, and we must not create a parochial biological survey that would tie the hands of the scientists and prevent them from pursuing their studies across American frontiers. This is why I have been advocating that it be a "national" rather than "U.S." biological survey.

Biological Survey Information: Introduction

Wallace A. Steffan
Idaho Museum of Natural History

Abstract: The sources of currently available biological data are discussed, e.g., publications, electronically stored data and museum collections, and the problems associated with integrating these potential data resources in a national biological survey. Various proposals are described for intra- and interdisciplinary standards for specimen or species-related documentation. Recent publications of relevance to a national biological survey are reviewed, especially those discussing the use of computers.
Keywords: Biological Data, Data Standards, Biological Survey, Museum Collections, Systematics Resources.

INTRODUCTION

Currently available biological data represent immense national resources. These are largely underutilized because of the difficulties in identifying these information resources, lack of uniformity in data documentation, and lack of sufficient national or interdisciplinary appreciation of the need for a biological resources information system. Armentano and Loucks (1979) described similar concerns in their evaluation of the national need for ecological and environmental data and the problems involved in integrating available data.

Attempts at standardization of data elements (there have been several) have not been widely accepted. For example, the report of the Association of Systematics Collections (ASC) Council on Standards of Systematics Collections (Black et al., 1975) was either not widely distributed or was largely ignored. Possibly the failure of this and earlier attempts to develop national standards for collection-related information was a result of the nature of the projects being proposed which were very ambitious but impractical at the time they were proposed because of a lack of adequate financial and professional support. I believe (and based on recent reports in other disciplines, others believe) that we now have the computer and human resources to plan and implement, realistically, a national biological survey. Education of and communication with all potential participants and with a significant cross-section of users of a national biological survey will be an essential prerequisite to the success of this endeavor.

SOURCES OF BIOLOGICAL SURVEY DATA

It would be presumptuous of me to go into any detail on the sources of biological data available for a national survey. I will attempt to outline very briefly some sources of data and some problems inherent in their utilization in a national biological survey.

There is no single bibliographic source of key works on the fauna and flora of the U.S. A publication similar to those published by the British [(Kerrick et al. (eds.), 1967; Sims and Hollis, 1980)] would certainly be a desirable initial phase of a U.S. biological survey and would reinforce the concept and utility of a national biological survey. A list of key references in systematics compiled by Knutson and Murphy (pers. comm.) could be expanded to provide such a bibliography.

One of the reasons for the failure of previous projects involving interdisciplinary collaboration may have been the lack of any useable product. Participants in or users of the services of a project of this scope need to see tangible results. Individuals within a particular discipline generally are not familiar with the literature in other disciplines. Publications such as "The Guide to the Literature of the Life Sciences" (Smith et al., 1980) are of some assistance in steering one in the right direction; however, review of the literature by specialists in each discipline would be an essential early phase in a national biological survey.

Published sources of information for biological survey data range from the latest catalogs of particular taxa, e.g., Diptera (Stone et al., 1965), *Mammal Species of the World* [Honacki et al. (eds.), 1982], to revisionary or monographic studies, to regional checklists, to descriptions of new species. Crovello's (1981) excellent discussion of the literature that serves as a rare plant information resource is certainly germane to the planning of a national biological survey. He discussed the problems with the literature as a data source and presented a set of recommended actions to enhance the value of this literature. The degree to which we develop an understanding of the literature as an interdisciplinary resource for a national biological survey will be a determining factor in its success.

The introduction of the personal computer has greatly accelerated the already burgeoning variety of data being stored electronically. The recent issue of the Federal Data Base Finder (Zarozny and Horner, 1985) identifies well over 3,000 data bases and files associated with federal activities. The Directory of Online Databases compiled and edited by Cuadra, et al. (1982) lists an equally rich and diverse selection of electronic data bases available for online searches through such information services as DIALOG. Neuner et al. (1981) compiled a computer file of names of vertebrates that is stored at the Association of Systematics Collections (ASC, Lawrence, Kansas). The extent of data bases at the regional, local and individual level and their potential value as data sources for a national biological survey are unknown but must be staggering.

The considerable current activity in the area of computerized catalogs or checklists would be directly relevant to a biological survey; e.g., *The Mammals of the World* and *The Amphibians of the World* (both published by ASC), *The Checklist of Nearctic Diptera* and the proposed *Biosystematic Database/Catalog of the Flies of the World* [both computerized files being compiled by the Biosystematics and Beneficial Insect Institute (BBII) of the U.S. Department of Agriculture], *The Catalog of the Hymenoptera of North America* (BBII and the Smithsonian Institution), and many other catalogs and national or regional lists in all disciplines. The duplication of effort along with "...inconsistencies in documentation of data, inadequate communication between data suppliers and data users, and a lack of overall coordination of the data bases in national research and monitoring programs..." (Loucks, 1985) represent a staggering waste of our financial and human resources.

Data in museum collections represent a massive information resource, but unfortunately most of this information is unavailable. The value of coming to grips with biological collection data was recognized years ago. Crovello (1967) discussed the problems in the use of electronic data processing in biological collections. Although technology and attitude towards computers has changed considerably since then, many of his concerns are still valid, especially those relating to problems associated with collections and curators. Barkley (1981) mentioned several factors of importance in assimilating herbarium label data, attributing the problems to intrinsic factors relating to individual specimen data and extrinsic factors relating to misinterpretations of distributional data through inference.

The complexity of museum or other systematics collections data varies according to the discipline involved. For example, collections of mammals or birds have relatively few species and specimens, but collections of insects may contain thousands of species and millions of specimens. These two types of collections need to be treated differently. Data on mammal collections can be specimen related; data on insect collections should probably be treated by lot (Humphrey and Clausen, 1977) or on an inventory basis (Steffan, 1985). As stated in the report, I firmly believe that systematists will remain the poor stepchildren of science until they realize the significance of the information content in collections as a whole, learn how to access this information, and, more importantly, how to make it available to others in the scientific and public communities. A national biological survey could certainly serve as a catalyst for this change in attitude and methodology, especially information management methodology.

The biological inventories, especially the species lists mentioned in "Data Management at Biological Field Stations" discussed below, could also be a major source of data for the survey. Many of the subprograms of the International Biological Program also produced computerized lists of biota investigated. Many other research programs of a local, regional, or national scope have also produced relevant data bases or files of electronic information.

The greatest potential source of new data for the biological survey is the expertise of individuals in all major disciplines. One of the first priorities for the planners of the survey would be to identify these human resources. The ASC has initiated electronic files of these human resources, and several disciplines have likewise identified these valuable resources. Human resources are on the one hand our greatest asset and on the other the major deficiency in implementing a national biological survey.

DOCUMENTATION STANDARDS

In their final report (Black et al., 1975), the ASC Council on Standards for Systematics Collections recommended six basic types of minimal data standards for all new biological collections and for all computerized specimen data banks. These documentation standards included:

1. The institutional identifier (acronym or number).

2. The item identifier (institutional subdivision and individual specimen catalog number).
3. The taxonomic identifier (phylum, class, order, family, genus, species).
4. The locality (continent or ocean; country or oceanic region; state, province, sea or major island group; county or other subdivision; latitude and longitude; and altitude in meters).
5. The time (expressed as year/month/day, e.g., 19850523).
6. The state of the specimen (part preserved, method of preservation, condition of item).

These recommendations have not been adopted universally; in fact, most potential users probably are unaware of them. The recommendations may have had more impact if they had been proposed in the detail presented in the documentation standards in mammalogy (Williams et al., 1979). Sarasan and Neuner (1983) highly recommended this compilation of documentation standards as an outstanding example of a data element dictionary; each data field recommended for inclusion in a mammalogy record is defined and acceptable formats described (see Table 1). The Canadian Heritage Information Network (CHIN) recently developed a similar natural sciences data dictionary (Delroy et al., 1985), which was published as a reference tool for users of their PARIS system. This data dictionary was based on recommendations of eleven subject area task forces, with representatives from the primary natural science disciplines, and additional fields recommended by end users and museum consultants. An example of the definition of each data field and associated descriptive comments is shown in Table 2.

With the exception of the CHIN documentation standards (designed primarily for museum collection data) no generally acceptable data documentation standards for integrated interdisciplinary data bases are available, although some of the interdisciplinary data bases described below have adopted their own standards.

Hierarchical coding systems for taxonomic catagories have been developed for a variety of interdisciplinary data bases; e.g., *A Taxonomic Code for the Biota of the Chesapeake Bay* (Swartz et al., 1972), the *National Oceanographic Data Center (NODC) Taxonomic Code* (1984), and the *BIOSTORET Master Species List* developed for the U.S. Environmental Protection Agency (Weber, 1976). The fourth edition of the NODC Taxonomic Code is available on magnetic tape, microfiche, and as a printed version.

RECENT PUBLICATIONS RELEVANT TO BIOLOGICAL SURVEY DATA

"Museum Collections and Computers"

This publication on museum collections and computers (Sarasan and Neuner, 1983) resulted from an attempt to review computerized management projects in museums. The purpose was to provide guidelines for museum curators and administrators considering such projects for their own institutions. The project itself was divided into three phases: 1) a mail survey of museum computer management projects, 2) site visits by Lenore Sarasan, the project leader, and 3) a review of the literature applicable to problems with museum data management.

The first four chapters of this study were based on the results of the mail survey

Table 1. Example of data category in mammalogy (from Williams *et al.* 1979).

Category: Type of preservation.

Description: This category applies to the mode of preservation of the specimen and all of its parts.

Format: Data entered in this category consist of a standardized two-character alphabetic code (see Valid examples).

Accepted variations: If the description of the specimen is not appropriate for any of the recognized codes then "OT*" (OT* = other) may be used. If this code is used or additional information is available for a specimen described by another code (example, skin and skull with supplemental histological and parasite preparations) an asterisk (*) should immediately follow the code to indicate that additional information is recorded in the category of REMARKS.

Omit conditions: This category has been declared mandatory for NIRM and is never omitted.

Contingency requirements: None.

Valid examples: The following codes represent NIRM standards for type of preservation:

Code	Definition
AL	Alcoholic
SS	Skin and skull
SB	Skin, skull, and partial skeleton
SN	Skeleton only (=all skeletal parts)
SK	Skull only
SO	Skin only
SA	Alcoholic and skull (removed)
KB	Skin and body skeleton
AN	Anatomical
PS	Partial skeleton
CO	Cranium only
HM	Head mount
BM	Body mount
SC	Skin, skull, and alcoholic carcass
BS	Body skeleton
OT*	Other

Other examples: SS*

Comments: Coding of data for this category promotes standardization of preservation descriptions and provides easier output operations that require this information, particularly output with limited working space.

and on interviews with museum project leaders. The major conclusion reached was that the success or failure of museum computer projects depended more upon decisions made by individuals planning and carrying out these projects than on the specific software or hardware used. Problems that museum administrators have encountered in computerization relate to inadequate project management, poor understanding of the principles and functions of documentation, and insufficient familiarity with the operation and application of computers. Four stages are recommended to avoid problems in setting up a computer information management system.

1. Preliminary research: study similar projects, conduct bibliographic research.
2. Determine needs: identify and analyze problems inherent in existing manual systems.
3. Systems analysis: develop a thorough understanding of structure and function of manual systems.

Table 2. Example of data category for PARIS system (see Delroy *et al.* 1985).

Sequence:	2550
Field label:	Specimen Nature
Field mnemonic:	SPENA
Field name:	Specimen Nature
Field definition:	This field describes the physical nature of the specimen
Entry rules:	Enter a keyword(s) to describe the specimen
Cataloguer's rules:	See also, Lot (LOT).
Data type:	Alpha-numeric string
Index class:	Phrase
Comments:	This field is in the National Database
Examples:	Study Skin
	Flat Skin
	Tanned Skin
	Mount
	Fluid
	Mummy
	Skull
	Skeleton
	Mandible
	Replica
	Model
	Artifact
	Slab
	Slide
	Mineral Specimen
	Loose Sediment
	Peel
	Thin Section
	Sidewall Core
Source:	Mammalogy; Paleontology; Invertebrate Zoology

4. Project goals: define concisely.

Once the problems have been identified, the needs assessed, and project goals established, the following stages need to be implemented:

1. System design: project director provides the system designer with a set of system criteria, including a description of what the system should do, what data should be maintained, and what outputs should be generated.
2. Develop a project plan and timetable.
3. Install the system.
4. Document the system.

The second part of this publication includes summaries for over 200 projects. Sixty-three of these may contain information relevant to a national biological survey. Major disciplines represented are botany, entomology, herpetology, ichthyology, mammalogy, and ornithology.

The third section, an annotated bibliography of computers and museums, lists the authors' selections of publications relevant to computerization in museums. The index provides a cross reference to access methods to computers, type of project, and software used.

This publication provides useful, although partially outdated, information for anyone planning to use computers in museums. The first four chapters should be read by anyone involved in planning information management systems for museums.

"Databases in Systematics"

"*Databases in Systematics*" (Allkin and Bisby eds., 1984) is the result of an international symposium on data bases in systematics, hosted in 1982 by the Systematics Association in England. It contains 26 papers presented at the symposium. As in the survey of museum computer projects (Sarasan and Neuner, 1983), it also reviews both the successes and difficulties of earlier projects. Although the emphasis of the symposium was on data bases to handle descriptive biological data, several papers dealt with general information management concepts of interest to planners of a national biological survey.

Heywood (1984) in "Electronic Data Processing in Taxonomy and Systematics," cogently discusses the lack of understanding of and commitment to information processing evidenced by most taxonomists. If a national biological survey is to succeed, this is one obstacle that must be resolved.

Bisby (1984) continues and expands this line of thought in his paper, "Information Services in Taxonomy." He stresses the service aspect of taxonomy, quoting Blackwelder (1967), "Its basic purpose is thus to systematize data for the use of other disciplines..." Bisby's "retail model" of the Taxonomic Information Service reflects this broad purpose of taxonomy. His "Diffusion Model" likewise reflects the relationships between taxonomic data and materials and the end user. He contends that both the taxonomic name and the descriptive biological data are essential attributes of an integrated taxonomic information system. Two of his major questions are: Who are the customers for the taxonomic information service? What do they want? He concludes that the introduction of data bases and modern communications technology provides an unprecedented opportunity to design and experiment with taxonomic information services of many styles and for the full range of potential customers.

Dadd and Kelly (1984), in their paper, "A Concept for a Machine-readable Taxonomic Reference File", describe a pilot project designed to provide the scientific community with an on-line, interactive tool for sharing taxonomic information via a computerized collection of organism names and associated data. This Taxonomic Reference File (TRF) will have four main components:

1. The Taxonomic Data File: includes organism names and nomenclatural data. Each name that has appeared in the literature will have its own entry and will be assigned a unique identifying number.
2. The Hierarchy File: contains classification schemes. It will carry multiple schemes to accommodate differences in classification.
3. The Related Data Files: includes multiple files containing different kinds of data, e.g. descriptive data about each organism, host/pest data, endangered species legislation.
4. The Bibliographic File: will serve as a link between entries in the TRF and the related bibliographic information found in existing printed and computer-readable products of BIOSIS.

The user will be able to copy desired portions of the TRF files into a separate work space area of the system. This type of application could be used by a national biological survey to initiate some of its files.

Heywood et al. (1984) describe the European Taxonomic, Floristic and Biosystematic Documentation System financed by the Research Councils of the 10 member countries. The first and current phase of this project is entry of the systematic, geographical, ecological, and chromosomal information cited in Flora Europaea (Tutin et al., 1964–80). Its aims are to provide a floristic, biosystematic, and taxonomic information system for the vascular plants of Europe. The first phase of the project has involved the analysis and review of existing relevant documentation systems. This research will allow the construction of a suitable file structure for the basic data in Flora Europaea and entry of those data into the system. The second phase will consist of updating the information in the fields covered by Flora Europaea from current and post-Flora Europaea literature, creation of new data field for information not included in this flora, and creation of a database suitable for various kinds of on-line searches.

The data base of the International Union for Conservation of Nature and Natural Resources (IUCN) Conservation Monitoring Centre (CMC) is described by Mackinder (1984) and deals with threatened animals and plants, protected areas, and wildlife trade. These areas of concern form the four main units of the CMC data base, which is set up to accommodate the two major criteria—taxonomic and geographic data—used to select and sort data from this type of information system. Both sets of data will be coded. For the U.S., the basic CMC geographic unit would be the state.

Flesness et al. (1984) describe "ISIS—An International Specimen Information System" with data on more than 100,000 animals held by zoological gardens and related facilities in twelve countries. Although not directly applicable to a national biological survey, the data standardization methods and coding used for the taxa are relevant. This paper is also relevant because it represents an information management system that functions across discipline, institutional and international lines.

Nimis et al. (1984) describe "The Network of Databanks for the Italian Flora and Vegetation". Data were standardized through the establishment of a central data bank. Nomenclature and coding were standardized for all data elements. Local data banks will be connected to form a networked data base.

"PRECIS—A Curatorial and Biogeographic System," described by Russell and Gonsalves (1984), is a data bank of herbarium specimen label information at the National Herbarium at Pretoria that includes data on the flora of South Africa. The data fields are either coded or freeform. Russell and Gonsalves present a detailed historical account of the development of the system and include definitions of problems encountered during the first year of development. The use of numeric codes for specimen identification, locality, and collector has resulted in a very high error rate and has affected the usefulness of the system. Despite this difficulty, progress in the development of an improved information management system has provided substantial results that could not have been attained otherwise.

Lucas (1984), in "Databases in Systematics: A Summing Up", concludes that systematists now have the tools to resolve what were impossible dreams a decade ago. He points to the coming of pragmatism, the appearance of microcomputers, and the availability of user-friendly software as three of the advances responsible for this change. All of these factors contributed to reaching the "critical mass" of involved and informed individuals essential to actual implementation of these projects.

"Data Management at Biological Field Stations"

This report, supported by the National Science Foundation and edited by Lauff (1982), addresses the need for data management planning at biological field stations. The workshop was organized around four general categories: 1) administration of data, 2) cataloging and documentation of data, 3) computers and software for data management, and 4) intersite exchange of information. The report should be consulted by individuals involved in data management planning for a national biological survey. In terms of availability of data for a national biological survey, the section on biological inventories, which includes species lists and collection indices, is directly applicable to this discussion.

The report is composed of an initial summary of recommendations followed by five chapters dealing with an overview of data management, databases, computer software systems, data administration, and exchange of information between sites. All sections are applicable to any planning process a national biological survey would initiate.

In its overview of data management, two perspectives are discussed, a research perspective and a perspective of secondary users, which would include the biological survey. The chapter on data bases presents discussions on data sets (encompassing data organization and structure, data coding, data entry, and record keeping), biological inventories (species lists and collection indices), documentation systems, data catalogs and directories, data banks, and integrated databases.

The chapter on computer software systems presents an excellent overview of the types of software recommended for data management and includes data entry, data dictionary, database management software, and integration of software systems.

Data administration, the topic of chapter 4, is a subject frequently neglected in the initial planning of large data systems in biology. The following recommended steps in establishing priorities are germane to the planning phase for a national biological survey.

1. Inventory of data bases currently planned or available.
2. Definition of the task and objective of each data set.
3. Prioritization of needs.
4. Determination of availability of resources.
5. Reassessment and reprioritization in terms of feasibility.
6. Selection of methods for completion of data management tasks.

This chapter also recommends criteria for selection of a computer system and discusses the need for data inventories, documentation procedures, and security.

The final chapter, on exhange of information between sites, succinctly discusses data exchange networks, protocol for exchange of data, mechanisms of exchange, and, very importantly, sharing of expertise on information management.

"Guidelines for Acquisition and Management of Biological Specimens"

This publication edited by Lee et al. (1982) represents the efforts of an interdisciplinary group of biologists charged with producing guidelines for the acquisition and management of biological specimens—especially voucher specimens. On the basis of the 1975 report of the ASC Council on Standards for Systematics Collections mentioned above, the participants of this Conference on Voucher Specimen Management formulated the following set of guidelines for mandatory categories of data that must accompany biological specimens if they are to be useful to investigators other than the collector.

1. A unique sample designation, possibly alphanumeric
2. The location of a sample collection site
3. Time and date of sample collection—as well as other biologically significant dates such as date of preservation, propagation, isolation, etc.
4. Name of collector—and other donor identification including station or field numbers
5. Identity to species level
6. Method of collection and preparation
7. Use of standard coding system
8. Distinct identifier for each repository

These guidelines are useful in a general way, but allow too much flexibility to be useful for a biological survey data base.

"Rare Plant Conservation: Geographical Data Organization"

This publication edited by Morse and Henifin (1981) was the result of a symposium sponsored by the U.S. National Park Service on synthesis of plant distribution information. It differs significantly from most symposium proceedings in that the publication reflects not only the discussions of the symposium, but also develops ideas presented at this meeting. The resulting 23 papers and nine appendices provide an excellent argument for cooperative projects such as a national biological survey.

Morse and Lawyer (1981) in their introduction state, "The lack of coordination and effective communication in the plant conservation field has resulted in much duplication of effort; some species and some geographical areas have been studied and reviewed repeatedly, and others not yet at all. Clearly the time has come to consider from many viewpoints the topic of Rare Plant Conservation: Geographical Data Organization, with particular emphasis on prospects for project coordination." The same ideas are applicable to a survey of our national biological resources.

Other sections of this book deal with information needs and priorities, information sources, descriptions of representative projects, and appendices treating topics related to rare plant conservation. Although the book covers only one discipline, botany, the ideas expressed and conclusions reached are certainly germane to any discussions of a national biological survey.

CONCLUSIONS AND RECOMMENDATIONS

Clearly some centralized mechanism needs to be established to begin to provide syntheses of information on the biological resources of the U.S. A national biological survey could serve such a vital function. Whichever options eventually are exercised (e.g., expansion of existing organizations or agencies or creation of a new organization), major problems involving synthesis of biological data from a variety of different sources and media need to be resolved. The Biological Survey of Canada (Terrestrial Arthropods) has gained momentum, and their publications (Danks, ed. 1982-) and expertise in development and implementation of a national survey of Canada should be a valuable asset in any planned national biological survey of the U.S.

The Canadian Heritage Information Network was developed after more than a decade of effort of numerous individuals in both humanities and natural sciences. The program's key mandates included development in information-sharing procedures and the creation of a national inventory of collections. Rottenberg (1984) presented an excellent overview of this program and is another valuable resource.

Several intitial steps could include:

1. Publication of a handbook to key references to the flora and fauna of the U.S. in a format similar to the "Bibliography of Key Works for the Identification of the British Flora and Fauna" (Kerrich et al. 1967). Smith et al. (1980) provide an excellent starting point for such a comprehensive publication.
2. Publication of information on human resources in each discipline following a format similar to that used in "The International Register of Specialists and Current Research in Plant Systematics" (Kiger et al. 1981).
3. Development of a data dictionary for the biological sciences in formats similar to those used by Williams et al. (1979) and Delroy et al. (1985). Foote (1977) provided a thesaurus of terms in entomology that would be one of the many sources to be used for an interdisciplinary data dictionary. Development of a data dictionary would involve a series of workshops–first at the disciplinary level and later at a national interdisciplinary level, at which some users and information management and documentation specialists would be involved.

LITERATURE CITED

Allkin, R. & F. A. Bisby (eds.). 1984. *Databases in systematics.* The Systematics Association Spec. Publ. No. 20. Academic Press, London. 329 p.

Armentano, T. V. & O. L. Loucks. 1979. *Ecological and environmental data as under-utilized national resources: Results of the TIE/ACCESS Program.* U.S. Department of Energy Contract No. EY-76-S-05-521. The Institute of Ecology (TIE), Indianapolis, Indiana.

Barkley, T. M. 1981. Use and abuse of specimen labels in distribution mapping. *In:* Morse, L. E. and M. S. Henifin (eds.). *Rare plant conservation: Geographical data organization.*

Bisby, F. A. 1984. Information services in taxonomy. *In*: Allkin, R. and F. A. Bisby (eds.) *Databases in systematics.*

Black, C. C. (ed.). 1975. Report of the ASC Council on Standards for Systematics Collections. *ASC Newsletter* 3(3): insert, 4 p.

Blackwelder, R. E. 1967. *Taxonomy: a text and reference book.* John Wiley and Sons, Inc., New York. 698 p.

Crovello, T. H. 1967. Problems in the use of electronic data processing in biological collection. *Taxon* 16: 483–491.

Crovello, T. H. 1981. The literature as a rare plant information resource. *In*: Morse, L. E. and M. S. Henifin (eds.) *Rare plant conservation: Geographical data organization.*

Cuadra, R. N., D. M. Abels & J. Wanger. 1982. *Directory of online databases.* Vol. 4(1): 1–293.

Dadd M. N. & M. C. Kelly. 1984. A concept of a machine-readable taxonomic reference file. *In*: Allkin, R. and F. A. Bisby (eds.) *Databases in systematics.*

Delroy, S. H., M. Cox, I. G. Sutherland & R. A. Bellamy. 1985. *Natural sciences data dictionary of the Canadian Heritage Information Network.* Documentation Research Publ. No. 2. 195 p.

Flesness, N. R., P. G. Garnatz & U. S. Seal. 1984. ISIS—An international specimen information system. *In*: Allkin, R. and F. A. Bisby (eds.) *Databases in systematics.*

Foote, R. H. 1977. *Thesaurus of entomology.* Entomological Society of America. College Park, Maryland. 188 p.

Heywood, V. H. 1984. Electronic data processing in taxonomy and systematics. *In*: Allkin, R. and F. A. Bisby (eds.) *Databases in systematics.*

Heywood, V. H., D. M. Moore, L. N. Derrick, K. A. Mitchell & J. van Scheepen. 1984. The European Taxonomic, Floristic and Biosystematic Documentation System—An Introduction. *In*: Allkin, R. and F. A. Bisby (eds.) *Databases in systematics.*

Honacki, J. H., K. E. Kinman & J. W. Koeppl (eds.). 1982. *Mammal species of the world.* Allen Press, Inc. and Association of Systematics Collections, Lawrence, Kansas.

Humphrey, P. S. & A. C. Clausen. 1977. *Automated cataloging for museum collections.* Association of Systematics Collections, Lawrence, Kansas. 79 p.

Kerrich, G. J., R. D. Meikle & N. Tebble (eds.). 1967. *Bibliography of key works for the identification of the British fauna and flora.* 3rd ed. Systematics Assoc. Publ. No. 1.

Kiger, R. W., T. C. Jacobsen & R. M. Lilly (eds.). 1981. *International register of specialists and current research in plant systematics.* Hunt Institute for Botanical Documentation, Carnegie-Mellon University.

Lauff, G. H. 1982. *Data management at biological field stations.* Report of a workshop held at the W. K. Kellogg Biological Station, Michigan State University. National Science Foundation, Washington, DC. 46 p.

Lee, W. L., B. M. Bell, & J. F. Sutton (eds.). 1982. *Guidelines for acquisition and management of biological specimens.* Association of Systematics Collections, Lawrence, Kansas. 42 p.

Loucks, O. L. 1986. Biological survey data bases: characteristics, structure, and management. *In*: Kim, K. C. and L. Knutson (eds). *Foundations for a National Biological Survey*, Association of Systematics Collections, Lawrence, Kansas.

Lucas, G. L. 1984. Databases in systematics: a summing up. *In*: Allkin, R. and F. A. Bisby (eds.) *Databases in systematics.*

Mackinder, D. C. 1984. The database of the IUCN Conservation Monitoring Centre. *In*: Allkin, R. and F. A. Bisby (eds.) *Databases in systematics.*

Morse, L. E. & J. I. Lawyer. 1981. Introduction. *In*: Morse, L. E. and M. S. Henifin (eds.) *Rare plant conservation: geographical data organization.*

Morse, L. E. & M. S. Henifin(eds.). 1981. *Rare Plant Conservation: Geographical Data Organization.* New York Botanical Garden, Bronx, New York. 377 p.

National Oceanographic Data Center. 1984. *NODC Taxonomic Code.* 4th edition. Key to Oceanographic Records Documentation Nr. 15. 2 vols., 738 p.

Neuner, A. M., T. J. Berger, D. E. Seibel, G. McGrath & D. Pakaluk. 1984. Checklist of vertebrate names of the United States, the U.S. Territories and Canada. Computer File, Association of Systematics Collection, Lawrence, Kansas.

Nimis, P. L., E. Feoli, & S. Pignatti. 1984. The network of databanks for the Italian flora and vegetation. *In*: Allkin, R. and F. A. Bisby (eds.) *Databases in systematics.*

Russell, G., E. Gibbs & P. Gonsalves. 1984. PRECIS—a curatorial and biogeographic system. *In*: Allkin, R. and F. A. Bisby (eds.) *Databases in systematics.*

Sarasan, L. & M. Neuner. 1983. *Museum collections and computers.* Association of Systematics Collections, Lawrence, Kansas. 292 p.

Sims, R. W. & D. Hollis (eds.). 1980. *Animal identification: a reference guide.* British Museum (Natural History), London. 198 p.

Smith, R. C., W. M. Reid & A. E. Luchsinger. 1980. *Smith's guide to the literature of the life sciences.* 9th Ed. Burgess Publishing Co., Minneapolis. 223 p.

Steffan, W. A. 1985. Inventory level computerization of systematics collections: an example for the Pacific. *In*: Sohmer, S. H. (ed.) *Forum on systematics resources in the Pacific.* Special Publications, Bishop Museum, Honolulu, Hawaii.

Stone, A., C. W. Sabrosky, W. W. Wirth, R. H. Foote, & J. R. Coulson (eds.). 1965. *A catalog of the Diptera of America north of Mexico.* U.S. Dep. Agric., Agric. Res. Serv., Agric. Handbook 276. 1696 p.

Tutin, T. G., V. H. Heywood, N. A. Burges, D. M. Moore, D. H. Valentine, S. M. Walters & D. A. Webb (eds.). 1964–1980. *Flora Europaea.* 5 vols. Cambridge University Press, Cambridge.

Williams, S. L., M. J. Smolen & A. A. Brigida. 1979. *Documentation standards for automatic data processing in mammalogy.* The Museum of Texas Tech University, Lubbock, Texas. 48 p.

Zarozny, S. & M. Horner. 1984. *The federal data base finder.* Information USA, Inc., Potomac, Maryland. 409 p.

BIOLOGICAL SURVEY DATA BASES: CHARACTERISTICS, STRUCTURE, AND MANAGEMENT

Orie L. Loucks

Holcomb Research Institute
Butler University

Abstract: Implicit in a national biological survey are large data files covering both the biota and the environment that sustains them. Also implicit is the need to have these data widely accessible to users for both basic research and for planning or policy assessments. Widespread use of biological data bases requires broad understanding of, and support for, high standards of data quality control, in the field, in the laboratory, in annotations and in computer summarization. This paper reviews 1) the options available to meet the immediate objective of more sophisticated data access and management of existing data through a national biological survey and 2) longer term goals of more complete data bases and more comprehensive syntheses of the compiled information.

Keywords: Data Bases, Data Management, Quality Assurance, User Groups, Syntheses.

INTRODUCTION

The characteristics and structure of the data bases comprising a national biological survey involve several loosely connected concerns: first, obviously, the biota and their environments; second, the skills and interests of those who study this biota; and finally, the practical concerns of a user community, ranging from students to researchers to developers to government and industry leaders. Two other considerations include where we are now in terms of the information base and the access of users to it and where we should be as a presumably sophisticated industrial nation building on and sustaining its renewable resource base.

Accordingly, my goals in this paper are 1) to review the characteristics of at least a major portion of the relevant data base as it exists now, 2) review problems of its management and user access, and 3) suggest approaches to a more comprehensive data management capable of meeting diverse user needs (including research) while also protecting proprietary interests in existing data.

PROBLEMS WITH THE STRUCTURE OF EXISTING DATA BASES

The presentations at this conference are replete with evidence as to the diversity of existing data bases covering museum and herbarium collections, plant and animal populations, life history data, and habitat characteristics. This situation is not at all new, and several studies have documented problems of access to, and reliability of, major biological and environmental data bases.

One of these studies was the ACCESS Project, supported in the late 1970s by the Department of Energy and performed by The Institute of Ecology (TIE) (Armentano and Loucks, 1979). The goal of ACCESS was to evaluate the national need for ecological and environmental data and determine the extent to which existing data documentation and archiving were meeting that need. The principal steps focused on then current data documentations and research in government, private, and academic sectors of the natural science community, particularly as they related to the accessibility of the data to secondary users.

The published results of the study, entitled "Ecological and Environmental Data as Under-Utilized National Resources," indicated that the potential contributions that existing biological data could make were not being achieved because of inconsistencies in data documentation, inadequate communication between data suppliers and data users, and a lack of overall coordination of the data bases in national research and monitoring programs. A nationally coordinated network was proposed, focusing on "regional data centers" and tied together through a hierarchy of data bases (national, state, and local) with a broad spectrum of potential users. A national biological survey could readily incorporate these hierarchically linked kinds of data (national, state and local) and thereby contribute to meeting important needs.

In evaluating the status of the existing data resources, the ACCESS project found that there are both gaps and duplications in the data bases throughout federal and state government, regional (inter- and intra-state) commissions, local agencies, academia, and industry. The reliability and associated documentation of these data bases varies widely. In some, such as certain of the state agency and industry air monitoring programs, there is a complete quality assurance record and careful control of data collection, analysis, and storage. For other data bases there is no written documentation and no systematic quality assurance program. Neither an awareness of these data as national resources nor the need for their accessiblity by other users has been recognized as an important priority.

In considering the problems of making the user community aware of and encouraging them to use existing data, the TIE study found that scientists, policy makers, and environmental assessment personnel sought data and information of many kinds, but often to no avail. In the single state (Texas) where a comprehensive data clearinghouse had been operating, existing data were collated from many sources (under state law) to meet needs from a broad spectrum of users. There already existed an extensive need for unsummarized, numerical data among industrial consultants, researchers, and government personnel, but in the absence of a service providing access to these data, the value of the data for policy-making and research went unrecognized. The TIE report found that many users wanted a mixture of data services, such as raw data plus reports of related work or bibliographic listings. Others wanted consultation with source personnel along with certain data. Both groups benefited in Texas from having a simple, effective mechanism for satisfying a wide range of data needs from a single source. A national biological survey or a network functioning hierarchically as a single facility could provide a smiliar service to the nation as a whole.

CHARACTERISTICS OF BIOLOGICAL SURVEY DATA

The most comprehensive survey of the broad spectrum of biological survey data bases is that edited by Lauff (1982) for the National Science Foundation and the Association of Biological Field Stations. The report notes the diversity of data collected at biological field stations, including maps, specimens, charts, field notes, microfiche, and computerized textual as well as numeric information. Some data, such as climate records, are of utility to a great number of researchers, while others may pertain only to processes or species at one field station. They note that the data include both long-term and short-term records, the former requiring long-term management to be useful. The data bases discussed in the Lauff report include data sets compiled by individual researchers for their own use as well as data developed by field stations for general use. They include not only data in the usual sense, but data bases of information about data, such as directories and catalogs of data (such as museum collections), and their associated documentation.

The Lauff report starts by noting that a data set carefully managed for its *primary* purpose will also be most useful to others. Although the originator of a data set will place priority on immediate data analysis needs, this use is not necessarily at odds with long-term data management and utilization goals. The documentation and management needed to make data available to secondary users are simply an extension of what researchers should do for their own purposes.

However, certain general principles can be applied to data characterization and management, no matter how complex the data set: Defining the types of entities about which there are data, and then developing the data about each type of entity. While many data sets are gathered by individual researchers or research teams for their own use, other data sets should exist as general resources at all museums, field stations, or study sites. Some of these data bases can be thought of as "biological inventories" that describe both species and ecological characteristics of the locale. Such data bases should be available as part of a national biological survey.

Two typical types of biological inventory—species lists and indexes to biological collections—illustrate some special data management needs, according to the Lauff report (1982). The species data often takes the form of printed lists arranged in a taxonomic or spatial sequence. Some lists are compiled by a researcher or instructor directly from observations; others are compiled indirectly from anecdotal data, published reports, or museum collections. In many respects, species lists can be managed just like any other data sets, but in cases where a species list is derived, in whole or in part, from other data, there are some additional data management issues. Such a list is, in effect, a summary of other data. A summary of data, by definition, does not include all the data from which it is derived. For this reason, it is best to maintain documentation on the link between summary species lists and their source data. Since biological communities are dynamic, species lists should be dynamic and reflect changes in distribution or nomenclature. Data bases are more easily updated when the species lists and the source material are computerized.

The second example, computerized indexes, increases the utility of biological collections by making it easier to locate specimens quickly and by making some of the data inherent in the collection available for efficient analysis. An index usually contains, for each specimen, data such as the taxonomic name, locality from which the specimen was obtained, name of collector, date collected, and other information describing characteristics of the specimen. Minimal data categories are reviewed in "Guidelines for Acquisition and Management of Biological Specimens" (Lee et al., 1982).

DOCUMENTATION SYSTEMS AS A PART OF DATA MANAGEMENT

The Lauff report also notes that documenting data is essentially an elaboration of existing practices (survey and measurement). In addition to documenting the scientific aspects of research, it is also necessary to document technical aspects of data handling, structure, and content. It is necessary that both scientific and technical documentation be available to secondary users, and it is desirable that they be handled in an integrated fashion. Primary users (contributors) and secondary users can deal more efficiently with data when its documentation follows a uniform format. Typically, there might be many data sets at a site, each data set consisting of one or more files (or tables). Each data set and each data file should be documented. In addition, each file will have several constituent variables, and some of these variables might be contained in more than one file. The variables also need to be documented. Thus, one should focus on three entities: data sets, data files, and variables. Tables 1, 2, and 3 (from Lauff, 1982) list the categories of documentation needed for data sets, data files, and variables, respectively.

The degree to which documentation is computerized will vary from site to site. Much of the documentation of variables, and some documentation of files, is handled more or less automatically by recent software. However, all software used to prepare computerized data bases should be fully documented.

No matter how sophisticated the technical aids, effective documentation for secondary users must be facilitated by appropriate administrative policies and procedures. Researchers, on their own initiative, may maintain documentation about data structure for their own use, given efficient tools for doing so, but documentation of the origin of their data tends to be left incomplete. The Lauff report recommends that a data management group review all documentation of data supplied by researchers for incorporation into a data bank to ensure that minimal standards have been met.

DATA BANKS AS PART OF BIOLOGICAL SURVEY DATA MANAGEMENT

Finally, the Lauff report (1982) and other documents (NAS, 1985) suggest that a data bank can be thought of as a data base of data bases. It provides researchers

Table 1. Categories of documentation for data sets.

1. Data set name	A name or code that uniquely identifies the data set.
2. Data set title	A title that describes the subject matter.
3. Data set files	A list of the data files that constitute the data set.
4. Research location	Information that identifies the site of the research or other project that generated the data.
5. Investigator	Names of the person(s) responsible for the research or other project that generated the data.
6. Other researchers	Names of other persons responsible for various phases of data collection or analysis, especially those who could conceivably be consulted regarding use of the data.
7. Contact person	Name of the person to contact for permission to use the data, and for help in locating and obtaining it.
8. Project	Description of the overall project of which this data set is a part (to place it in the context of other research and to describe its purpose).
9. Source of funding	
10. Methods	Description of methods used to collect and analyze the data, including the experimental design, field and laboratory methods, and computational algorithms (via reference to specialized software where necessary). (This category is analogous to the methods and materials section of published papers. It could easily be subdivided into other categories. The experimental design, especially, could be put in a separate category, since it can help describe the rationale of the data set.)
11. Storage location and medium	Storage location and medium of the data set as a whole, e.g., magnetic tape, disk files, punched cards, etc.
12. Data collection time period	A description of the data collection period and periodicity, and major temporal gaps or anomalies in the data set pattern.
13. Voucher material	Site (institution, collection) where voucher material has been deposited.
14. Processing and revision history	A description of data verification and error checking procedures, and of any revisions since publication of the data.
15. Usage history	References to published and unpublished reports or analyses of the data that could be of interest to a secondary user.

with a single source for all data pertaining to a site and can ensure a degree of quality and consistency in the management of data and documentation. Most of the work needed to develop and maintain a data bank pertains to the ways in which data are entered into it. Although the development of storage structures and search tools (the "output" system) for use by secondary users is an important task, it is even more important to develop methods for obtaining cooperation and data from contributing researchers (the "input" system). In the absence of automated techniques for dealing with the problem of data updating and documentation, the Lauff report recommends the establishment of a regular system of review. Each data set and its documentation should be scheduled for periodic review by the contributing researcher, who can be requested to note any updates or corrections that should be applied to the data or documentation. The period between reviews can be short when the data set is relatively active and relatively long (on the order of years) thereafter.

Control of data quality is another concern. Quality control and quality assurance mean different things to different people. Aspects of quality control in biological surveys range from the scientific to the technical. They include the quality of research (e.g., quality of hypotheses and experimental design), quality of mea-

Table 2. Categories of documentation for data files.

1. File name	A name or code that uniquely identifies the file.
2. Constituent variables	A list of the variables contained in the file. This list (and the information about each variable, i.e., the categories listed in Table 3) is the most important information about the file.
3. Key variables	A list of the hierarchy of variables that determine the sorted sequence of the data, or a list of the variables that constitute the file's "key."
4. Subject	An explicit description of the subject matter of the file. It should make clear what type of entity is described by the records.
5. Storage location	A description of the location of the file (in terms of a computer system's file naming system, where appropriate).
6. Physical size	The number of records and total number of characters, or other such descriptors.
7. File creation methods	A description or list of procedures or algorithms used to create the file, and the files from which the file was derived (if applicable).
8. Update history	A record of updates to the file (where those records might help to reconcile differences with previous versions of the data).
9. Summary statistics	A brief set of summary statistics (means, sums, minima, maxima, etc.) for each variable. (These can be used to verify that the data file one is using is indeed the correct version, and to verify the accuracy of data transfers.)

Table 3. Categories of documentation for data variables.

1. Variable name	The name of the variable (which should be unique within the data set), and any synonyms which a user might encounter.
2. Definition	A definition of the variable in ecological terms.
3. Units of measurement	
4. Precision of measurement	(Statements about precision should not only give error bounds, but explain what they refer to. The user should know whether the variance given is that of determinations by an instrument, or among replicate samples at a single location, or among locations within a given area, etc.)
5. Range or list of values	The minimum and maximum values, or for categorical variables, a list of the possible values (or a reference to a file that lists them and any code definitions).
6. Data type	A description of the variable, in terms like "integer," "date," "4-byte real," or whatever others are used by a data base management system (DBMS) or statistical package. (This information is needed when dealing with data stored in the special formats of a DBMS or statistical package.)
7. Position and/or format	Any information that will be needed by a program in order to read data from (for example) an ASCII file. (This information is typically needed in a non-DBMS environment and is almost always needed for data transfer between sites.)
8. Missing data codes	A list of codes that indicate missing data. If there are several types of missing data codes, they should be distinguished.
9. Computational method	Algorithms that were used to derive this variable from others (if applicable).

surement (e.g., adequacy of instrumentation and methods, replication, confidence limits), and quality of recording and transcription of data (e.g., from field forms to computer). In a sense, these terms simply extend existing understandings that the quality of research and of measurements can be controlled in large part through documentation of methods and the data obtained. If all data are thoroughly documented as to persons responsible, methods, etc., a later user can decide whether a particular data set is of sufficient quality for a particular purpose.

Quality control in the area of data recording and transcription is particularly troublesome. Data entry is one level that is prone to error. Much time is wasted when errors are found in data at advanced stages of analysis, requiring correction and reanalysis of entire studies. Even worse from a scientific standpoint, poor quality control may leave errors undetected until years later. Whatever data verification procedures are followed, the documentation should make them clear to the user.

DATA BASE MANAGEMENT FOR BIOLOGICAL SURVEY DATA

The report by Lauff (1982) notes that data base management systems (DBMSs) are the most general and basic of data handling software. Different people will have different ideas of what they are because the meaning of "data base management system" often depends on whether it is used in the context of mainframe computers or microcomputers. Persons who work with business data bases on large computers would not consider the DBMSs available for microcomputers to be worthy of the name, while for a person operating in a microcomputer environment, the DBMSs used on large computers are unnecessarily complex and more of a hindrance than a help to accomplishing useful work. A DBMS, if comprehensive enough, can tie all other software and data together by serving as a general purpose storage and retrieval system for all types of data. A common data structure can make possible a consistent treatment for all data. Tools for error checking, documentation, and security are easy to develop and to use if the data are in a common form. A DBMS can also include a language for retrieving and manipulating data. These two features, a generic structure for data and a set of generic operations to manipulate data, can free the researcher from many of the details involved in performing the same functions in general purpose programming languages (Lauff, 1982).

Data can, of course, be managed without the software that goes under the name "data base management system." Sometimes other software products, alone or in combination, provide some of the functions that we might otherwise obtain from a DBMS. We consider here three important features from the Lauff report:

Generic data structure—A uniform data storage structure can do much to integrate data management. It is far too confusing and wasteful to have to store data in one way for one analysis and in another way for others. A good DBMS will make it possible to store all data in a uniform way, yet retrieve them easily in the form required by any other software.

Data independence—A DBMS can make the data storage structures independent from the programs that use the data. This makes it possible to change a data base without disrupting programs that use it. A good DBMS, however, will make many types of changes possible without necessitating changes in the programs that read or write the data.

Security control—A DBMS can control access to data by allowing the manager of a data base to make specified portions of it available to certain persons, for specific purposes (e.g., updating, reading).

AN "ECOSYSTEMS DATA HANDBOOK" AS A PRODUCT OF A NATIONAL BIOLOGICAL SURVEY

As modern science has grown, specialties within it have matured into separate disciplines. Typically, communication within each new discipline becomes more and more circumscribed; at the same time, various activites within the discipline differentiate and become new sub-disciplines. Ecosystem science as a discipline has coalesced from the several disparate disciplines (including species distribution and community data) that contributed to its foundation. Thus, the data bases required for ecosystem analysis should be recognized as part of a national biological survey. Several papers in this volume have cited the need for some "overarching synthesis" from the survey. One such synthesis would be application of the data to improving ecosystem management.

Although the literature of ecosystem science has been accumulating for almost a century, it was not until the onset of the International Biological Program (IBP) that much systems-oriented ecological material was gathered. The advent of the National Environmental Policy Act of 1969 (NEPA), the Council on Environmental Quality (CEQ), and the U.S. Environmental Protection Agency (EPA) led first to recognition of the inherent interdependencies of man and natural-resource systems, secondly to new data about ecosystems, and thirdly to a pressing need to apply those data.

Accordingly, late in 1974, CEQ convened a meeting at the Smithsonian Institution in Washington, D.C., where prospective users of ecosystem data examined the difficulties of obtaining and applying biological and ecosystems data to societal problems. The participants concluded that the ready availability of information about ecosystems in the form of a handbook would facilitate greatly the preparation of environmental assessments and the solution of resource-use problems (TIE, 1979).

Two general assessments were made of the availability of data. One involved an inventory of relatively automated information sources as compiled by personnel at the Oak Ridge National Laboratory. Information was also obtained from The Encyclopedia of Information Systems and Services (Kruzas and Schnitzer, 1971). Finally, information was used from preliminary results of a survey of computerized data bases in the Midwest conducted by staff of TIE's ACCESS Project. The survey resulted in identification of some 208 sources of data or information about data, as of January 1977. Only seven of the sources listed an estimate of the number of observations contained in their data bases; these averaged 9,740,000 observations each. Some 21 other data bases were identified from the survey.

Numerous problems concerning the format and scope of an ecosystems handbook emerged, however. One problem centered on the comprehensiveness of the handbook's contents and difficulties involved in obtaining a consensus as to what should be included. Also of great importance was the need to document the data base so that the handbook contents would meet a broad range of user needs.

The following are the criteria agreed upon to limit the scope of an ecosystems handbook (TIE, 1979):

- ⋆ The handbook should not significantly overlap with existing physical and biological handbooks.
- ⋆ The geographical area for which data are needed is North America (north of Mexico), associated island ecosystems (including Hawaii and Puerto Rico), and inshore marine ecosystems.
- ⋆ Handbook data should relate to the biotic characteristics and processes of natural resource ecosystems, excluding, however, both human ecology (except perturbations caused by human activites) and intensive production cultures (such as agriculture and aquaculture), except as these activites might act as stressors on natural ecosystems.
- ⋆ Introductory (or appendix) material should provide guidelines for users, sources of information, summaries of ecosystem concepts and principles, literature aids, and a glossary.

Information concerning the content of the handbook was obtained from two surveys, a user survey and a content survey (to 454 respondents). From the user survey, certain types of topical needs were identified (see Table 4). Since the study group recognized that all the necessary background information about the data could not be conveyed in the individual chapters, there would be a need for a narrative introductory chapter. The introduction itself would describe the scope of the data to be summarized. The organization of the information in the handbook also would be discussed briefly to explain how the biomes and ecosystems had been identified.

The introductory section also could provide guidelines for users of the data tables. These guidelines would cover:

- ⋆ Comparability of Methods and Data. The users of the handbook would be made explicitly aware that in recent years the methodology of obtaining ecological information has changed in some areas (e.g., standing-crop and productivity). Such information has been obtained by clipping quadrats as well as by determining the evolution of CO_2. This example illustrates why users must be fully informed about how the data are collected.
- ⋆ Specific Uses and Scope of Data. The handbook would be useful on four principal levels:
  - + Policy analysis and formulation related to environmental impact, larger scale spatial/temporal planning, and resource use/management models
  - + Decision-making on policy options, by various government and private units
  - + Ecosystems resource management
  - + Research and instruction
- ⋆ Uses in Policy Analysis and Decision-making. Because of the importance of

Table 4. Proposed table of contents for the ecosystems data handbook.

Title of volume (unit)	Chapter subheadings*
1.0 Introduction and ecosystem principles	Scope of the handbook
	Organization of handbook criteria
	Criteria for biome and ecosystem divisions
	Table of biomes and inclusive ecosystems
	Ecosystem description and data classes
	Map of biomes
	Sources of information used for data
	Guidelines for users
	Mcthods and data comparability
	Specific data uses
	Scope
	Uses in policy analysis and decision making
	Dangers of misapplication
	Ecosystem components, attributes, and inter-relationships
	Ecosystem cybernetics
	Principles of ecosystem energetics
	Biogeochemical flows and cycles
	Ecosystem responses and recovery from disturbances
	Aquatic ecosystem commonalities across biomes
	Terrestrial ecosystem commonalities across biomes
	Land-water interactions
2.0 Grassland biome	Shortgrass ecosystem
	Tallgrass ecosystem
	Mixedgrass ecosystem
	Palouse prairie ecosystem
	Desert grassland ecosystem
	Annual grassland ecosystem
	Mountain grassland ecosystem
	Everglade grassland ecosystem
	Appropriate aquatic ecosystems
3.0 Desert biome	Great Basin desert ecosystem
	Mojave desert ecosystem
	Chihuahuan desert ecosystem
	Sonoran desert ecosystem
	Appropriate aquatic ecosystems
4.0 Broad-schlerophyll biome	Oak Woodlands ecosystem
	Chaparral ecosystem
	Appropriate aquatic ecosystems
5.0 Pinyon-Juniper biome	Pinyon-Juniper ecosystem
	Appropriate aquatic ecosystems
6.0 Tundra biome	Tall shrub ecosystem
	Low shrub ecosystem
	Cottongrass ecosystem
	Graminoid ecosystem
	Polar desert ecosystem
	Alpine desert ecosystem
	Appropriate aquatic ecosystems
7.0 Boreal-taiga biome	Shrubland ecosystem
	Appropriate aquatic ecosystem

Table 4. Continued.

Title of volume (unit)	Chapter subheadings*
8.0 Temperate coniferous forest biome	Sierra/Cascade Montane ecosystem Rocky Mountain Montane ecosystem Appalachian Montane ecosystem Northern Pacific Coast ecosystem
9.0 Temperate deciduous forest biome	Oak-Chestnut Forest ecosystem Oak-Hickory Forest ecosystem Mixed Mesophytic (Forest) ecosystem Western Mesophytic Forest ecosystem Southeast Evergreen Forest ecosystem Beech-Maple Forest ecosystem Maple-Basswood Forest ecosystem Hemlock-White Pine-Northern Hardwoods Forest ecosystem Appropriate aquatic ecosystems
10.0 Island biome	Island forest ecosystem Island Montane ecosystem Island Grassland/March ecosystem Coral reef ecosystem Appropriate aquatic ecosystems
11.0 Subtropical biome	Terrestrial ecosystems unspecified Appropriate aquatic ecosystems
12.0 Coastal biome	Near-shore marine ecosystem Estuarine ecosystem Seagrass and algal ecosystem Marsh ecosystem Mangrove ecosystem Dune ecosystem Barrier Island ecosystems
13.0 Ecological literature and organizations	Ecological literature aids Ecological review journals Research reference services Related handbooks Journals Organizations primarily concerned with Ecology Professional and educational societies Research organizations Conservation and protection organizations

* A glossary, constants, coefficients, conversions, literature citations, and an index will be incorporated in each volume focusing on data for one or more biomes.

biological data in societal policy making and decision-making processes, and because of the relative underdevelopment of this interface, the narrative should include a brief but illustrative discussion of:

+ The relationships among biological data, other information, and various factors used in policy analyses
+ The process of transforming and synthesizing these data for policy and decision-making purposes

- ⋆ Dangers of Misapplication. The guidelines would also address the danger of misuse of the handbook data, as in situations where a data point may be valid but should not be used in interpretation and/or application for certain concerns. Included would be the danger of releasing pinpoint sites for endangered or threatened organisms.

Within each of the proposed biome volumes, a series of ecosystems will be listed for complete summarization of data. For each type of ecosystem, information would be provided for purposes of comparison. The overall outline of the content of the handbook (Table 4) should be regarded as tentative. Because the availability of data varies widely among the proposed biome volumes, the correspondence between "biome" and published "volumes" may need to be reassessed at a later date. At the same time, however, the outline recognizes that additional data are becoming available. Future updates could lead to much more complete treatments.

Also to be incorporated into the handbook are data needed for the development of ecological models to aid in understanding ecosystems. Literature aids, referrals to state-of-the-art reviews, and a glossary of terms are included in the final table of contents, in response to a high level of interest expressed by potential users.

FINDINGS AND RECOMMENDATIONS

1. A very large quantity of biological data (taxonomic, demographic, and life-history) and associated environmental description is available nationally, but access is fragmented, and very serious data gaps exist regionally and taxonomically.
2. Over the shortterm, existing data bases should be viewed as underutilized national resources. Existing data in museum collections, at biological field stations, in ecological research institutes, and in large-scale or long-term university-based ecological research should be treated as an information resource to be made accessible in the first years of a national biological survey to a national audience of users.
3. Data management systems should be planned so as to benefit both the primary research and secondary users of the data bases. The sometimes conflicting viewpoints of different types of users and institutions should be reconciled so that data management practices complement each other.
4. One of the long-term products of a national biological survey should be a synthesis of such data, possibly in the form of an Ecosystem Data Handbook.

LITERATURE CITED

Armentano, T. V. & O. L. Loucks. 1979. *Ecological and environmental data as under-utilized national resources: results of the TIE/ACCESS program.* U.S. Department of Energy Contract No. EY-76-S-05- 5213. The Institute of Ecology TIE, Indianapolis, Indiana. 98 p.

Kruzas, A. T. & A. E. Schnitzer (eds.). 1971. *Encyclopedia of information systems and services,* 1st ed. Edwards Brothers, Ann Arbor, Michigan. 1105 p.

Lauff, G. H. 1982. *Data management at biological field stations.* Report of a workshop held at the W. K. Kellogg Biological Station, Michigan State University. National Science Foundation, Washington, DC. 46 p.

Lee, W. L., B. M. Bell, & J. F. Sutton. 1982. *Guidelines for acquisition and management of biological specimens.* Association of Systematics Collections, Lawrence, Kansas. 42 p.

National Research Council. 1982. *Data management and computation. Volume 1: issues and recommendations.* Committee on Data Management and Computation, Space Science Board. National Academy Press, Washington, DC. 147 p.

The Institute of Ecology. 1979. *Feasibility study report on a proposed Ecosystems Data Handbook.* Report to the National Science Foundation. (Grant DEB 75-20525.) TIE, Indianapolis, Indiana. 70 p.

Development of Research Information Systems: Concepts and Practice[1]

Melvin I. Dyer
Michael P. Farrell
Oak Ridge National Laboratory

Abstract: This paper provides a review of system-oriented constraints on the development of a nationally integrated data base to serve the needs for a national biological survey. Such an effort must take cognizance of the classical "Tragedy of the Commons" scenario to avoid critical pitfalls. An important new awareness is developing around the subject of hierarchical ordering in biological and environmental systems, and this subject needs to be considered in depth in the development of a national biological survey. Development of specific programs to attend to problems of uncertainty and aggregation in the construction and use of data sets is particularly important. We present recommendations for a two-tiered system to assemble and disseminate information for a National Biological System. One tier is for the individual user, and the other is for use prompted by large, centralized programs funded by any government or private agency that has a specific mandate to address.
Keywords: Research Data Management Organization, User Orientation, "Tragedy of the Commons," System Hierarchy, Uncertainty Analysis, Problems In Aggregation.

INTRODUCTION

Many groups stating an inherent interest in a U.S. national biological survey are represented in this volume: academicians interested in assuring that valuable biological and systematics data are cared for; state scientists and administrators wanting to ensure that biological and environmental information is available for the well-being of their constituencies; federal scientists and administrators charged with large and wide-ranging mandates for resources in the whole of the U.S.; and private industry scientists and administrators interested in a variety of problems ranging from environmental concerns to those with more local or immediate needs in industry. Added to this group are those from outside the U.S. who have

[1]Research sponsored in part by the Carbon Dioxide Research Division, U.S. Department of Energy, under Contract No. DE-AC05-840R21400 with Martin Marietta Energy Systems, Inc.

expressed their points of view. This wide scope places extreme demands on any system that purports to assemble and disseminate information on the scale called for by most authorities stating their opinions here. That a national biological survey would be an excellent idea when implemented is not questioned; rather, what is consistently emphasized is the plea to assemble a useful and trustworthy system. That, of course, is a daunting task in view of experiences with environmental data assemblages. What are some of the technological and theoretical limitations that we must regard for development of a national biological survey, and what then are some of the known ways that technological solutions can be applied to solve some of the difficulties?

DEVELOPMENTAL PHILOSOPHY

The "Tragedy of the Commons" and Research Data Banks

Garrett Hardin's classic paper on the "Tragedy of the Commons" (Hardin, 1968) publicized events following attempts to structure activities about a commons. Those same lessons hold for Research Data Management (RDM).

A commons is defined as anything that can be utilized or shared together by a number of individuals. Many examples exist, but the one most easily understood was originally put forward by Hardin (1968, 1974), that of grazing lands, not individually owned, where individual herders graze their cattle. As long as the carrying capacity of the commons is kept in order, there is no difficulty. However, therein lies the potential for the tragedy, the essence of unfolding dramas. Each individual seeks to maximize his or her gain, thereby incrementally stressing the system. As long as the system is infinitely robust, there will be no difficulty, but, as we all know, systems are bounded and finite. Thus, there will come a point where the original system will cease to function because its capacity for responding positively is exceeded. As long as use of the commons operates without external controls it can never be self-correcting. The reason is simple. Hardin showed that benefits accruing to the individual by incrementally increasing use are $+1$, whereas the negative utility for that individual is only a fraction of -1. Therefore, the gains of cheating to the individual are always greater than are any deficits so long as the system is capable of sustaining the increased demands. It is when everyone applies this principle that the system collapses, to the detriment of everyone.

How does this fit with the development of RDM? Interestingly, it fits in two ways: 1) from the standpoint of entry of information, and 2) from the use of information stored in files. Both affect the quality of the information emanating from the overall use of the entire RDM program. We must also consider how inputs and outputs are structured when information is gathered from the scientific community at large and how this structuring affects programs designed to develop and acquire information that are funded for special purposes.

The first way that the Tragedy of the Commons can affect RDM is in the initial effort to acquire data. For example, while working on the International Biological Program, one of us (MID) was active in a task for assembling what was purported to be "all" of the information collected by the Grassland Biome Project for incorporation into a data bank. The data bank was needed to synthesize information gathered during this experimental entreé into what was billed as "big biology" (Hammond, 1972). In attempting to pull together the information from many sites over many years, it was necessary for MID to contact the scientists who had contributed to the Grassland Biome Project in an attempt to acquire their data sets. Published papers, analyzed but unpublished data, and sometimes raw data were involved. From the outset, central program staff sought to incorporate quality control into the data bank development. Several methods were proposed to achieve this aim, but eventually each investigator simply was asked to certify that the information he or she was providing was correct and accurate to within a certain level of error. Great resistance was encountered from the investigators, particularly in regard to data for which no peer-reviewed papers had been published. Ultimately the grand design for a massive data bank containing all Grassland Biome Information collapsed, partially because it was not possible to certify the accuracy of the information.

How does this story relate to the commons? It does so through the fact that each individual operating as he or she chose as a responsible scientist could not be made to feel that they were a part of the commons unless their own interests were served first. The positive component was either a publication of the data sets with their names, or the assurance that the information was indeed the best they could provide. Many investigators could not provide assurances of quality control because of their need to publish the information, and moreover, that tackling the job was worthwhile. Thus the peer-review process is deeply embedded in building a commons for RDM purposes (Marzolf and Dyer, 1986). We have no statistical data for this variation of the tragedy of the commons in environmental sciences and ecology today, but we suspect that it happens often. It has backlash, as was exhibited during a National Science Foundation forum held in 1980 to review the second round of recommendations related to Experimental Ecological Reserves. Remarks were brought forward at the end of that meeting about quality control not being assured until full peer-reviewed publication. Frustration about such apparent need for publication was expressed by agency managers in one person's statement: "We simply cannot wait for the pages of *Ecological Monographs* to appear before our decisions can be made." Doubtless that statement is true. However, if it is equally true that the most prudent way for information to become acceptable by all in the commons is through the peer-review process, then the question remains, do we have much of a choice? We can hastily assemble data and information for emergency measures and hope they are acceptable, or we can develop a quality-assured information base, one with a greater chance of withstanding the scrutiny of time, by the slower route—obtaining data from research appearing in the commons through the well-established and accepted peer-review process. The main difference is that published data, while not infallible, probably are the most cost-effective source over the long term. The open literature pathway is slower, and its total data assemblage undoubtedly smaller, than initially inexpensive but hastily prepared surveys. However, the published record is more often than not the authority. This, too, is part of the Tragedy of the Commons.

The second way that the Tragedy of the Commons can affect RDM is through the use of data in centralized data banks. The data bank is itself the commons. For logistical and economic reasons, such a commons probably exists in only limited form. This means that access to most data banks is limited and probably cannot be made available to all potential users in a society, much the same way that not everyone who wants to can graze cattle on western rangeland. Similarly the users who can access the data bank have constraints dictated by the commons. Because it is costly to organize and run a data bank, some sort of financing system must be developed. Although an organization sometimes assumes all financial obligations, usually the standard is a fee system in which costs are allocated to each potential user, most often based on anticipated use and estimated ability to pay. Often a data bank is subsidized in a number of ways. If that is the case, then the true meaning of the Tragedy of the Commons becomes fully apparent. If those who are most wealthy feel they are subsidizing the operation by paying the largest fees, they may tend to feel they can monopolize the data bank resources much the same way that the largest cattle owner might tend to monopolize the best grazing lands or those containing the water sources. Those who can afford lesser units of usage, but who still have responsibilities or requirements that cause them to spend time on the system, will tend to use it on increasing incremental bases. The ultimate fate of such developments is the tendency for RDM or computer systems to become plugged with longer and longer queues, which become increasingly unacceptable to the individual user. At some point the capacity of the system is exceeded, followed by a variety of changes that ultimately may result in major disruption of the RDM or even bankruptcy of the entire activity. Because a problem now exists in the entire operation, effort is dedicated toward finding a solution, most often some sort of technological solution. Depending upon the nature of the problem and the type of solution, such technological changes may or may not be welcome or even useful. Often the solution imposed is the creation of new RDM centers or offshoots, developments that may not always be successful since we now have lost the central facilities that seemed to make the RDM commons needed in the first place. Hardin (1968) warned that not every problem necessarily has a technological solution. This well may be the case for certain types of RDM.

The "Catch 22" Problem

Another problem has emerged in conceptual form in the past few years that must be addressed in the development of a national biological survey. We said in the introduction that many persons attended this conference, each with their own concept of what a national biological survey should contain. These concepts range from thoughtful comments about the scientific needs for such a survey (Kim and Knutson, 1986), catalogs listing diversity and genetic diversity in natural systems (Steffan, 1986; Schonewald-Cox, 1986), and the state of systematics collections (Chernoff, 1986) to agricultural ecosystems (Johnson, 1986; Klassen, 1986) to considerations of enormous natural systems (Loucks, 1986), conservation matters (Jenkins, 1986), environmental protection (Hirsch, 1986), and finally a combination of many of these (Risser, 1986).

This is obviously a complex set of interests. How do we organize the survey to make it meaningful, yet simple enough to use and robust enough to persuade all but the most recalcitrant potential users of its utility? The task is not easy. For instance, critique of the development of the rather massive undertaking by the

National Science Foundation in its Long-term Ecological Research Program was centered about what the chosen sites should consider collecting for the long-term record, and what should be the underlying bases for collecting such information. In the sixth year of the research program, it is still not altogether clear in some instances how to address these seemingly simple questions. A new question is now being posed: Is it possible to address ecological problems of the future when information being collected today does not have those future questions to drive the collection process? Certainly we all know that published data or pre-existing but unreported data sets are sometimes valuable for problems we encounter today, and we are delighted when that comes about. But, we argue, this is truly serendipity that we accept after the fact. Thus, at the same time we ask whether scientists as representatives of society can responsibly construct a program on *ipso facto* bases, solely for the sake of hoping that someone someday will be able to use the information. This places us in a "double bind" (a condition identified by Bateson as discussed by Hardin, 1968), which is somewhat related to the idiomatic expression of "Catch 22" introduced by Heller (1955). If we do not know what the question is, how can we collect information about it? Our way out of this problem has been to use overall experience and generalizations to structure new programs. But, if we choose wrongly, can we then address the now unknown questions that almost certainly will emerge later? This is the first part of the "Catch 22." Since we recognize this as a paradox, our first reaction is to turn to technology for the answer, regardless of whether it helps us or not. Here our technological solution is "then let's collect everything we can, hoping that someday it will be useful." This action activates the second phase of the "Catch 22." Can we possibly afford to "collect information about everything," and, if we can, will we be able to sort the wheat from the chaff when the problems of the future arise for which we were purportedly laying plans? The probable answer is sometimes yes, but, unfortunately, more often no. Nonetheless, we must alway keep in mind the continuous task of incorporating older data sets into the RDM.

This point brings us full circle in planning for a national biological survey. We submit that we are almost carrying out this scenario today in modern biological and ecological science. We have literally thousands of research staff in universities and laboratories in the U.S. and elsewhere collecting information, only part of which ultimately become placed in the public record. Once again we have demonstrated a facet of Hardin's Tragedy of the Commons.

NEW PERSPECTIVES

The Hierarchical Nature of Biological and Ecological Systems

To this point we have painted a rather cynical picture of the development of a national biological survey system which, ironically, we fully support. How might we avoid some of the commons tragedies and problems with scenarios where we introduce technology as a solution, where with a bit of thought we can realize that there are no technological solutions? The first subject we might address is that of how the information ought to be organized. There is increasing awareness that biological and ecological associations exist in a hierarchical manner, and it follows that information (data or models about them) should be organized in a similar manner (Allen and Starr, 1983; Allen et al., 1984). We cannot give a recipe for the formulation, but we foresee that this approach to the problem of building a reasonably interactive and accessible information base with very large and disparate entities will pay off. The types of information called for by those interested in systematics collections can be made to fit this scheme easily because they are hierarchical in the first place. Those systems with greater amounts of complexity, such as the types that Loucks (1986) has reviewed, can also be made more tractable by hierarchical ordering. Only in this way can the enormous task of constructing the national biological survey system be accomplished.

Analysis of Uncertainty and Understanding of Problems in Aggregation

Before giving our attention to the framework we advocate for developing the national biological survey, we turn to one last subject, which is academic now, but in the future will have to be regarded routinely for almost every use of any data assemblage, such as that being considered for the national biological survey, particularly if information collected from small-scale endeavors is aggregated into large-scale syntheses.

Expression of component error is an inherent part of our modern biological and ecological world, thanks to decades-long development of mathematical and statistical methods. But as we leave observations of individuals and turn our attention toward groupings or aggregations of observation, we must invoke models of the world from which we extract individual measurements. We have had to reorganize our thoughts about uncertainty in cases in which data—and models for expressing the meaning of those data—are used (Gardner, in press). Instead of having discrete units to assemble from a data source (nature or data banks), we now must consider assemblages or rate processes, graphic plots of which are often statistically smoothed out or skewed. Thus it is not a simple task to show cause and effect relationships in answer to many biological and ecological questions. To date these techniques have been applied to a variety of environmental problems: radiological assessment (Hoffman and Gardner, 1983; O'Neill et al., 1981), generic problems dealt with by ecological models (O'Neill and Gardner, 1979), assessment models (Downing et al., 1985), stream ecosystem models (Gardner et al., 1981), marsh hydrology model (Gardner et al., 1980a), global carbon models (Gardner et al., 1980b), and predator-prey models (Gardner et al., 1980c). In all instances, information that included data from data banks was used along with a variety of models to provide an in-depth assessment of a particular problem. Even though much emphasis was placed on ways to assess uncertainty in modeling studies, the lesson is clear. Studies involving models cannot be divorced from data, particularly during validation and use of those models for planning or developing management alternatives.

Once it has been decided to invoke information from a data bank, it is necessary to account for discrepancies that might be there. A simple example is the extrapolation of information from point sources to a spatially large scale. This is often needed, particularly for environmental concerns, because, for economic reasons, seldom are studies repeated uniformly over large expanses of space, or over all time periods necessary to represent the problem at hand. Coupled with the uncertainty analyses is the question of how one should aggregate information to avoid problems with error (Gardner et al., 1982). Because complexity forces ecologists to deal with aggregates at some point, some attention must be given to the way error affects the ability of an investigator to assemble meaningful project descriptions. This will be especially important to the development of a national biological survey as it brings together a program in which data and models must be used for an assessment.

Lastly, as exercises involving models and data are performed, what should their fates be? Are those results not then equally good candidates for inclusion into the national biological survey?

RESEARCH DATA MANAGEMENT SCHEMA

The intent for a Research Data Management program is manifold. It must satisfy the overall requirements first set globally, and it should satisfy the needs of any individual who wants to utilize it. For a national biological survey this is a daunting requirement. However, we assert, the underpinnings of such a system lie in ensuring that the individual user can be satisfied. The example cited earlier about the U.S. IBP experience is instructive. If the individual is unfettered by superfluous administrative and management restraints and unencumbered by extraordinarily complex hardware and software systems, then, in our estimation, the system will work. Personal computer hardware technology and sophisticated software development have combined in recent years to release us from having to create the Tragedy of the Commons in the development of a national biological survey. Once the overall intent, scope, and approach are established and agreed upon, then the fate of the project will ultimately reside in the hands of the individual scientists who can work in close association with the administrators of the program.

Having said that, we now want to progress to the design of a project that has served well thus far in instances of government and industry where these design criteria have been applied. The first such system design we will address has programmatic components of a very general nature to help the individual user. The design has as its central focus the needs of the research and development community where specific problems require extensive data bank entities, but the problems may be more localized or of a smaller scale, such as an examination of floral or faunal characteristics of the country or the development of descriptions of highly specific or long-term phenomena. The second system design addresses truly large biological or environmental problems, such as global carbon problems or acidic deposition, where the magnitude of the problem requires well-defined group dynamics of a large organization. The first design more often addresses the needs of the individual investigator or large groups of small teams, whereas the second design addresses governmental or private agencies with special tasks and where large numbers of investigators in relatively few work forces are assembled.

Research Data Management for Small Programs

The individual investigator must decide whether his or her research projects will benefit from a formal Research Data Management program. This decision will probably be made on the basis of the type and volume of information being collected and the overall complexity of the project. For those investigators, many options are open, ranging from simple programs they may develop themselves to sophisticated commercial spreadsheet programs now available for use on personal computers.

For larger integrated programs where several investigators collaborate or for specific organizations such as biological stations, there is a growing awareness of the need for the development of a centralized data and information management system. The most recent review of protocols, concepts, and everyday usage was presented at a symposium of data management for the National Science Foundation's Long-term Ecological Research Program (Michener and Marozas, 1986). Many of the papers published in the proceedings of this symposium dealt with aiding the individual investigator who is attached to either interdisciplinary or multidisciplinary projects or who works at a national or state facility. The subject is discussed in considerable depth in the symposium proceedings, and we will not give additional coverage here. Individuals and projects that follow the precepts presented in that publication can avoid pitfalls we discussed in the first section of this paper.

How Do We Start: A Case History

"One of the most important components of this information flow is the timely exchange of accurate, usable data. Thus, the data coordination function becomes one of the most important areas that must be addressed within the National Acid Precipitation Assessment Program (NAPAP)" (Benkovitz and Farrell, 1983). Substitute NBS (National Biological Survey) for NAPAP and the statement describes the situation faced by the development of a national biological survey: What should we do about RDM needs? The solution for the NAPAP was to have a Data Coordination Plan drafted, reviewed by the participants, updated/changed, and, finally, implemented. The NAPAP plan identified two levels of coordination needed in managing the data flow within the program. A first level of coordination was set to define the substantive contents of the needed data; these included variables to be measured, geographic and temporal coverage, minimum level of disaggregation, quality assurance procedures, etc. A second level of coordination was needed to define data-access techniques, possible transfer formats, software selection, etc. To implement these levels of coordination, two complementary activites were implemented: (1) creation of the Data Coordination Core Activity (DCCA) within the Interagency Task Force (ITF), whose job it is to ensure the fulfillment of the methodology and data requirements of the assessment activities, and (2) the expansion of the DCCA structure to include the design and implementation of the data management support facilities to carry out integrated assignments. This two-phased approach by the ITF appears to have been very successful. In fact, a group established at Oak Ridge National Laboratory to implement most of the activities has been a valuable resource to the NAPAP when new data-intensive tasks are needed (e.g., Lake Water Quality Survey).

Obviously, the above case history demonstrates the need for drafting a data coordination plan for a national biological survey, plans for which could be very similar to, or very different from, the NAPAP plan; we do not know which at this time. What is clear is that the Tragedy of the Commons' scenario outlined previously has an unfortunately better chance of being on-target without a data coordination plan as compared to starting the national biological survey program with an implementation plan that can serve as a temporary model.

The Information Analysis Center

If the Data Coordination Plan is successful in identifying the practical aspects of information and data flow within a program, then a group must be identified to handle specific tasks such as development of RDM techniques and scientific and technical information (STI) exchanges. However, there is more to the problem than putting staff together to handle the various tasks. Can this group enhance the information/data so that a "product" from the data is improved, i.e. more complete in documentation, evaluated/certified, error-free, etc.? The answer is yes, but do not look for many groups spanning the spectrum of activities needed by large research programs such as the national biological survey. The lack of broad-based information groups is not because of the inability to identify the need but rather the lack of funding. Adding value to information is expensive and is not often recognized as critical to the success of a program. However, those programs with identified and funded information analysis centers have a direct, tangible benefit to be gained from adopting this information enhancement model.

For example, the Carbon Dioxide Information Center (CDIC) of Oak Ridge National Laboratory, sponsored by the U.S. Department of Energy (DOE), Carbon Dioxide Research Division, maintains, summarizes, and distributes information resulting from CO_2 research worldwide. The nature of CDIC's charter dictated that CDIC be more than an information clearinghouse—more than an organization that collects and disseminates "information about information" concerning CO_2. A "value added" concept was adopted that permitted CDIC to process, assure quality, document, disseminate, and evaluate CO_2-related information. It was decided that CDIC would follow the "information analysis center" model characterized by staff trained in various scientific disciplines related to DOE's research areas (climatology, botany, terrestrial and aquatic ecology, mathematics, oceanography, etc.), the computer sciences, and various information science specialties (bibliographic systems, information extraction, information searches, etc.). Hence, CDIC's multi-talented technical staff maintains an information analysis center that interacts with the CO_2 research community worldwide and DOE's programmatic components on a broad scientific spectrum, performing original analysis of extant data, evaluating scientific reports (as opposed to abstracting information), and testing and documenting complex computer models.

We believe that the information analysis center concept is critical to the success of a national biological survey. As with the CO_2 issue and CDIC's role in that program, a national biological survey is also a long-term project that must integrate STI from multidisciplinary sources. The prospect of adding value to the STI is overwhelmingly supportive of establishing a national biological survey information analysis center. In fact, we feel that establishing such a center is an immediate obligation of NBS program management.

PROBLEMS IN APPLYING RESEARCH DATA MANAGEMENT TO THE NATIONAL BIOLOGICAL SURVEY

Creation of a National Biological Survey Information Analysis Center would solve most of the problems of coordinating activities among investigators, research programs and various agencies. However, creating a RDM structure that is responsive and appropriate for the national biological survey will not be a trivial task. It is true that many software tools exist to bridge the identified needs, but the philosophy of RDM is currently heavily influenced by the business environment. In this section we wish to point out our experiences working in the research environment and elaborate on our ideas of what an RDM system should be and how to ensure that the RDM structure does not hinder the science behind the program.

The General Problem

The foundation of sound research data base (RDB) management is detailed planning (Martin, 1976). One who manages such a data base is usually barraged by the apparent need for flowcharts, PERT diagrams, cross-reference libraries, directional dictionaries, and a host of other "aids" designed to increase efficiency and/or ensure that the final product will accomplish the project goals. Plans for recovering data sets, sorting strategies, merging and updating capabilities, and accomplishing intercomputer exchanges must be planned well in advance with few allowances for "lurking variables" (Box et al., 1979). The research data manager must also give detailed breakdowns on development time, personnel and computer costs, and the lead time necessary to develop the application programs even though the project may be several years away with the possibility of personnel turnover and hardware changes.

Informal polls among research data managers indicate that an increasing number of research data management systems are not being developed as outlined in many of the leading references in this field. For example, flowcharting, one of the backbones of the industry, has been shown to be an academic exercise with little application or help to real world complex data base problems. PERT diagrams are very informative *after* the final report is written. Dictionaries, fixed sorting strategies, cumbersome merging capabilities, data set recovery problems, etc., are no longer problem areas because of software developments.

Furthermore, most research data managers find it difficult, if not impossible, to cost-justify more than a cursory research data management plan. Time estimates for development are usually too long and must be reduced. In addition, cost estimates may be very inaccurate when many research programs cannot identify all the variables or data base formats that will ultimately be necessary for analysis and report generation.

Despite the previously mentioned problems in research data management applications, a discernable naivety regarding economic feasibility is associated with many of the currently published reports dealing with RDM. One of the principal reasons may be the different approaches in research data management found among applied contract researchers with strict budgetary constraints and university-based research programs with their more liberal approach to computer-related costs. Although free computer time at universities is becoming rare, there still exists such a price differential as to perhaps subliminally encourage many of the data management strategies popularized in reference material.

Another area of concern in RDM is the problem (most obvious among research data managers trained as programmers) of viewing most projects as unique with unique solutions. The list of in-house-developed data base management programs written in FORTRAN and/or COBOL specifically for a project must be prodigious. Our own experiences could supply a long list of data base management structures so specific as to preclude any general use and often meeting only a small proportion of the needs of the RDB manager because of cost overruns, programming problems, or changes in the project's emphasis. On the other hand, this tendency to "reinvent the wheel" does not seem as popular among research data managers who have been trained in areas other than programming and who are aware of the new application programs currently available on lease or purchase options.

The National Biological Survey and Research Data Management

Unlike many other projects, a national biological survey will probably have most of its problems identified previously, with project goals often defined more appropriately after the study becomes operational. Many times at the start of a project, variable selection and research data formats are tentative because of the unknown biological complexity that may be encountered. Potential ways to summarize the data base are usually more numerous than money permits. Lead time for development of even a simple RDM structure is usually nonexistent. Research data managers frequently become involved with a project only shortly before data collection, with the subsequent need for immediate data summarization, so that the project may be modified before the next scheduled sampling period. In such an atmosphere, where the research data managers face a project that will provide answers by an iterative process and in which there will be major changes in the data base content and structure, the manager cannot hope to spend many days planning the specifics of the RDM, and only infrequently can the cost of development of such an RDM be justified.

Faced with the uncertainties of managing a national biological survey data base, a research data manager is expected to provide a project with a skeleton of a data management system that can add broad ranges of new variables, reformat existing variables, perform unanticipated analyses; provide computer-generated tables and copy-ready figures that will be formatted at the conclusion of the study, and, in general, produce immediate answers via a time-sharing system but provide the capability to reduce the cost of large, complex analyses via batch operations. Superimposed on the above list, the system development must not demand excessively large and complex programming tasks. The system must be cost-effective, with costs ranging from 5 to 10 percent of the total project funds. Furthermore, the need for a system analyst to manage and construct the data base is, by project definition, counterproductive. The requirement to have a research data manager is often considered an unnecessary burden to the project and may jeopardize the project's financial capability to measure other important variables or eliminate some other aspect of the project. If the project group includes a statistician, it is usually that person who is nominated to manage the data base.

The System Selection

Although the usual decision as to the choice of appropriate software is often based on hardware availability, an appropriate software system should be selected unencumbered by the normal hardware consideration. The projected software system for a national biological survey should meet the following five major system selection criteria: 1) have vendor support of the system's software including programming applications, analysis programs, and help in troubleshooting user applications; 2) provide not only easily-programmed, user-oriented, flexible, and hierarchical data management capabilities with canned instructions (e.g., sorting, merging, updating), but also user-programmed instructions (e.g., input, output, quality control checking); 3) provide a basic complement of statistical analysis routines (i.e., means, standard deviation, analysis of variance, regression, etc.), plotting and charting capabilities, and more advanced programs that may be available in the system, programmable within the system, or available in other packages that interface with the parent system; 4) provide a common syntax for batch and time-shared operation; and 5) be cost-effective, not only in terms of computer costs [e.g., core, central processing unit (CPU), input/output devices] but also in terms of the personnel time needed for implementation and maintenance.

Planning Versus Open-Ended Management

All data management structures must be planned. What is perhaps not clear is the amount and direction of planning necessary after an advanced analysis package has been selected. Detailed planning of research data bases appears to be inversely proportional to the degree to which the selected package meets the system selection criteria outlined previously. If adherence to the criteria is high, then planning the RDM system can be minimal, with sessions devoted to determining output formats and requirements and any specialized analysis programs needed but not contained in the package. On the other hand, when a selected package has major omissions or when adherence is low, in regard to the system selection criteria, planning time is usually increased with more emphasis being placed on the basic problems of research data management such as variable input formats, internal file construction, sorting, merging, and updating. Therefore, adherence to the system selection criteria permits the research data manager to be more involved with the end-product requirements of the study, such as copy-ready graphical displays, computer-generated tables, and quality control assurances. In turn, the scientist involved with reporting the findings of the study benefits from this new end-product orientation of the research data manager. Now the scientists can become more involved with interpreting results of the study than with editing mandated by an ineffective system. Furthermore, decisions that were previously based on inflexible computer programs can be modified to place the emphasis on the scientist's needs. As a result, efficiency is gained in field operations, where the majority of cost is usually involved, without additional cost to the data management program.

The term "open-ended research data base management" (OE/RDM) has been coined to describe this philosophy of data management that supports a minimal planning effort, one in which the emphasis is placed on computer-related needs at the completion of the study (Farrell et al., 1979). Obviously, an OE/RDM strategy will not work for all types of programs, nor for all levels of experience in executing the selected package. In addition, OE/RDM may not be appropriate as a working system because of program emphasis and/or the degree to which the selected package meets the system-selection criteria.

We feel that a national biological survey will need to implement an OE/RDM to meet the challenging and changing nature of the program.

SUMMARY

A national biological survey without an effective research data manager embedded in an information analysis center that is the result of a tactical and strategic plan will probably fail to meet the expectations of the scientific community. If such a plan is not implemented, we predict that the Tragedy of the Commons will, once again, claim another noble experiment. It seems that we are in a position with the development of a national biological survey to avoid playing out the tragedy. This conference is a first start towards a real solution of how to initiate a national biological survey. We hope our comments are taken as positive and constructive, not as pessimistic or stultifying. As we stated, we support the development of a national biological survey program and hope that some of our comments will serve to help build the model for this program.

LITERATURE CITED

Allen, T. F. H. & T. B. Starr. 1982. *Hierarchy: perspective for ecological complexity.* The University of Chicago Press. 310 p.

Allen, T. F. H., R. V. O'Neill, & T. W. Hoekstra. 1984. *Interlevel relations in ecological research and management: some working principles from hierarchy theory.* USDA For. Serv. Gen. Tech. Report RM-10, Fort Collins, Colorado. 11 p.

Benkovitz, C. M. & M. P. Farrell. 1983. *Data coordination program plan for the interagency task force on acid precipitation.* Brookhaven National Laboratory Report No. 51845, 34 p.

Box, G. E. P., S. Hunter, & W. Hunter. 1979. *Statistics for experimenters.* John Wiley and Sons, New York. 653 p.

Chernoff, B. 1986. Systematics and long-range ecologic research. *In*: Kim, K. C. and L. Knutson (eds.). *Foundations for a national biological survey.* Association of Systematics Collections, Lawrence, Kansas.

Downing, D. J., R. H. Gardner, & F. O. Hoffman. 1985. An examination of response-surface methodologies for uncertainty analysis in assessment models. *Technometrics* 27: 151–163.

Farrell, M. P., A. D. Mayoun, & K. Daniels. 1979. Management of evolving ecological data sets with SAS: An open ended management approach. *In*: *Proceedings of the Fourth SAS Users Group,* International SAS Institute, Cary, North Carolina. 453 p.

Gardner, R. H. In press. Error analysis and sensitivity analysis in ecology. *In*: M. Singh (ed.). *Encyclopedia of systems and control.* Pergamon Press, London.

Gardner, R. H., D. H. Huff, R. V. O'Neill, J. G. Mankin, J. Carney, & J. Jones. 1980a. Application of error analysis to a marsh hydrology model. *Water Resources Res.* 16: 659–664.

Gardner, R. H., J. B. Mankin, & W. R. Emanuel. 1980b. A comparison of three carbon models. *Ecological Modelling* 8: 313–332.

Gardner, R. H., R. V. O'Neill, J. B. Mankin, & D. Kumar. 1980c. Comparative error analysis of six predator-prey models. *Ecology* 61: 323–332.

Gardner, R. H., R. V. O'Neill, J. B. Mankin, & J. H. Carney. 1981. A comparison of sensitivity analysis and error analysis based on a stream ecosystem model. *Ecological Modelling* 12: 173–190.

Gardner, R. H., W. G. Cale, & R. V. O'Neill. 1982. Robust analysis of aggregation error. *Ecology* 63: 1771–1779.

Hammond, A. L. 1972. Ecosystem analysis: biome approach to environmental research. *Science* 175: 46–48.

Hardin, G. 1968. The tragedy of the commons. *Science* 162: 1243–1248.

Hardin, G. 1974. Living on a lifeboat. *BioScience* 24: 561–568.

Heller, J. 1955. *Catch-22.* Dell Publishing Company, New York. 463 p.

Hirsch, A. 1986. The role of a national biological survey in environmental protection. *In*: Kim, K. C. and L. Knutson (eds.). *Foundations for a national biological survey.* Association of Systematics Collections, Lawrence, Kansas.

Hoffman, F. O. & R. H. Gardner. 1983. Evaluation of uncertainties in environmental radiological assessment models. *In*: J. E. Till and H. R. Meyer (eds.). *Radiological assessment: a textbook on environmental dose assessment.* U.S. Nuclear Regulatory Commission, Washington, DC. NUREG/CR-3332, ORNL-5968.

Jenkins, R. E. 1986. Applications and use of biological survey data. *In*: Kim, K. C. and L. Knutson (eds.). *Foundations for a national biological survey.* Association of Systematics Collections, Lawrence, Kansas.

Johnson, R. L. 1986. Plant protection and a national biological survey. *In*: Kim, K. C. and L. Knutson (eds.). *Foundations for a national biological survey.* Association of Systematics Collections, Lawrence, Kansas.

Kim, K. C. & L. Knutson. 1986. Scientific bases for a national biological survey. *In*: Kim, K. C. and L. Knutson (eds.). *Foundations for a national biological survey.* Association of Systematics Collections, Lawrence, Kansas.

Klassen, W. 1986. Agricultural research: the importance of a national biological survey in environmental protection. *In*: Kim, K. C. and L. Knutson (eds.). *Foundations for a national biological survey.* Association of Systematics Collections, Lawrence, Kansas.

Loucks, O. L. 1986. Database structure and management. *In*: Kim, K. C. and L. Knutson (eds.). *Foundations for a national biological survey.* Association of Systematics Collections, Lawrence, Kansas.

Martin, J. 1976. *Principles of data-base management.* Prentice-Hall, Englewood Cliffs, New Jersey. 352 p.

Marzof, G. R. & M. I. Dyer. 1986. Summary and future directions of research data management in ecology. *In*: W. Michener and M. Marozas (eds.). *Research data management in the ecological sciences.* Proc. Symposium Nov. 4–7, 1984, Hobcaw Barony, Georgetown, South Carolina, University of South Carolina Press.

Michener, W. & M. Marozas. 1986. *Research data management in the ecological sciences.* Proc. Symposium Nov. 4–7, 1984, Hobcaw Barony, Georgetown, South Carolina. University of South Carolina Press.

O'Neill, R. V. & R. H. Gardner. 1979. Sources of uncertainty in ecological models. *In*: B. P. Zeigler, M. S. Elzas, G. J. Kliv, and T. I. Oren (eds.) *Methodology in systems modelling and simulation.* North Holland Publishing Co.

O'Neill, R. V., R. H. Gardner, F. O. Hoffman, & G. Schwarz. 1981. Parameter uncertainty and estimated radiological dose to man from atmospheric ^{131}I releases: A Monte Carlo approach. *Health Physics* 40: 760–764.

Risser, P. G. 1986. State and private legislative and historical perspectives. *In*: Kim, K. C. and L. Knutson (eds.). *Foundations for a national biological survey.* Association of Systematics Collections, Lawrence, Kansas.

Schonewald-Cox, C. 1986. Diversity, germplasm and natural resources. *In*: Kim, K. C. and L. Knutson (eds.). *Foundations for a national biological survey.* Association of Systematics Collections, Lawrence, Kansas.

Steffan, W. A. 1986. Biological survey data: introduction. *In*: Kim, K. C. and L. Knutson (eds.). *Foundations for a national biological survey.* Association of Systematics Collections, Lawrence, Kansas.

Public and Scientific Dissemination of National Biological Survey Data

Nancy R. Morin
Missouri Botanical Garden

Abstract: The primary medium by which information will be disseminated from a national biological survey will be hardcopy publication. As computer technology advances, information may be disseminated through diskettes, tapes, and online access, which also are considered here to be publications that might result from a national biological survey. A national biological survey could make previously produced, relevant publications more accessible and could serve to coordinate production of new publications under an easily indexed and retrievable single series. By canvassing existing literature and available specialists, a national biological survey could identify specific areas in which publications are urgently needed and could encourage and facilitate work on such publications. Among the kinds of publications that might be produced under the auspices of a national biological survey are systematic surveys of major groups on a national basis, checklists of species, monographs of taxa, and special reports on selected subsets or attributes of organisms (e.g., life forms, chromosome numbers, distribution). The steps taken in the Flora of North America project provide an example of those that might be taken by a national biological survey toward production of other urgently needed publications. These steps are to define the need, assess resources, identify potential users, and design a workable organization for completion of the project. A maximally useful national biological survey should coordinate such efforts, sponsor or facilitate publication, and help disseminate the resulting information.

Keywords: Checklists, Floras, Flora of North America Project, Monographs, Publications.

INTRODUCTION

The purpose of this paper is to consider the kinds of publications that might result from a national biological survey. These publications are likely to be primarily systematic in nature, but if a national biological survey also were to embrace other disciplines many other kinds of publications might result as well.

The primary medium by which information will be disseminated from a national biological survey will be hardcopy publication. Although some major users of national biological survey-generated information may have on-line electronic access to the information, I believe that the expense and complexity of maintaining this capability will prohibit the majority of users from having such access. An increasing number of people may prefer to have the information stored and available on diskettes for use in personal computers, but storage of data on a diskette is simply a variation on the book (hardcopy) theme and does not truly represent a substantially different medium.

Most of the examples below relate to botany but are general enough to apply to most groups of organisms. The publications mentioned are precisely the kinds that have resulted from the research performed by systematic biologists through the years or that systematic biologists have recognized as being needed in their discipline. There is nothing novel about the kinds of publications that could be expected from a national biological survey.

Why Have a National Biological Survey?

What contribution would a national biological survey make, then, if biologists are already producing the kinds of publications that are needed? First, these publications would become much more readily visible and available, especially to biologists outside the subject discipline, because all publications could be associated with one easily indexed and retrievable series. In preparing this paper, I searched several large biology libraries and concluded that it is extremely difficult and time-consuming to locate information if one is not already familiar with the literature of a particular discipline. Through this activity a national biological survey would thus provide the scientific community with a mechanism to assure that such information is made accessible to all users, thereby facilitating interdisciplinary collaboration and helping to prevent unintentional duplication of research effort.

Second is the question of need. Are biologists already producing all of the publications needed? The answer is certainly no. But because of a lack of coordination among researchers on various taxa, the scientific community is unable to consider all organisms when ranking its needs. The first priority of a national biological survey should thus be to determine which publications are most urgently needed, after assessing the available resources. Needed publications already in progress through individual or group effort outside of a national biological survey might receive additional logistical or financial support from a national biological survey. A national biological survey would provide the motivation and coordination required to produce needed publications for which expertise is currently available. In those cases in which a publication is needed but there are no specialists capable of producing it, a national biological survey might be instrumental in developing training programs or in encouraging universities to offer such programs.

KINDS OF PUBLICATIONS

Resources

The first publication generated by a national biological survey might appropriately be a guide to collections, specialists, and institutions of general interest to biologists. Each of us in our own discipline probably has some kind of directory—perhaps produced by a professional society. For example, the "International

Register of Specialists and Current Research in Plant Systematics" (Kiger et al., 1981) and "Index Herbariorum I" (Holmgren et al., 1981) were compiled on the basis of responses to extensive questionnaires, and both are used frequently by botanists. Similar directories would be useful for other groups, with specialists indexed alphabetically or by geographical location and by taxon and geographical area of expertise and with collections indexed by taxon and geographical area. A national biological survey could provide a mechanism to assure that such information is made generally available and easily accessible to users—including biologists in other disciplines. For example, such accessibility would assist an agricultural extension officer trying to determine where to send for identification specimens of a bramble suspected to be new to an area.

Complementary to publication of a listing of specialists would be a publication of bibliographic assessment of the available literature. An example of such a publication is the series "Research Service Bibliographies," produced by the State Library of South Australia (e.g., Lovett, 1970). Many bibliographies certainly will be published as a result of and to assist in the work of a national biological survey. Many such bibliographies now exist, but the question again is how accessible are they to potential users and which additional bibliographies are needed most urgently for researchers to proceed efficiently with their work?

Information

Moving from a discussion about publications on information resources to the information itself, we find that a national biological survey would make two major, unique contributions. First, on the basis of our knowledge of available resources and after an assessment of urgency, a national biological survey should help us recognize specific areas exhibiting the greatest needs for publications. When these needs have been identified, a national biological survey can encourage and provide funding for the production of those publications. Second, where a treatment of some kind is needed on a national scale, a national biological survey would facilitate the coordination and funding of such a project. In examining the kinds of publications that might be considered, I first examined those produced by other national biological surveys, then analyzed those which had been produced by other kinds of organizations or individuals. I concluded that our imaginations may be the only limit to the kinds of documents that might be published. What must be asked in each case is whether such a publication is needed, how urgent it is, and if such a publication is already being produced by individuals or institutions.

Most prominent among national biological surveys are systematic surveys of major groups, such as a flora or fauna. The Australian Bureau of Flora and Fauna is producing a *Flora of Australia*, for instance. The first volume contains general information about the Australian flora and a key to the families of flowering plants of Australia (George, 1981). Subsequently published volumes treat the families and include keys, descriptions, pertinent literature citations, maps, critical illustrations, and lists of representative specimens. This multi-volume work will not serve as a field guide, but it was identified as the most urgently needed work for this extremely diverse and interesting flora. A similar project is underway for the flowering plants of southern African through the efforts of the Department of

Agricultural Technical Services, Republic of South Africa. The authors have produced keys to and descriptions of the genera of southern African flowering plants (Dyer, 1975) and are publishing synoptical treatments of families (Codd et al., 1966-).

Variations on publications covering major taxa on a national basis range from checklists of species for entire areas (e.g., Jessop, 1983), on the one hand through comprehensive monographic treatments of all groups in an area on the other (e.g., Komarov et al., 1934-64).

To date, the information provided in these major works has not been saved in a computerized database. A program called the European Documentation System was recently instituted by V. H. Heywood (Heywood, unpub.) to enter the information contained in *Flora Europaea* (Tutin et al., 1976) in a computerized database. This information will be available online to users and will allow publication of reports on numerous subsets of the information contained in this important and massive flora. Development of a similar database through a national biological survey would not only facilitate production of such reports on the U.S. biota but would also allow direct comparison between at least the floral aspect of the biota of Europe and that of the U.S.

Works that cover selected subsets or attributes of the biota of an area include a variety of topics. One subset includes regional works, such as *The Flora of Central Australia* (Jessop, 1981), published by the Australian Systematic Botany Society with government financing. A second subset treats life forms, for example *The Handbook of Native Trees of South Australia*, produced by their Woods and Forests Department (Boomsma, 1981). Still another subset includes specialized information about all of the species in a region, for instance the *Flora Europaea Check-list and Chromosome Index* (Moore, 1982), such as could be generated easily from a complete national biological survey database and updated frequently. Similarly, a comprehensive atlas of the distribution of groups, such as that published on the European flora (Jalas and Suominen, 1983), could be generated from information in a national biological survey database. Another subset could treat parts or life stages of organisms, such as the desperately needed keys to the identification of different life stages of insects. For plants, these publications might include atlases of woods and seeds (Berggren, 1903-) and pollen (Cerceau-Larrival, 1980). Lists of and additional information about rare plants might also be produced (e.g., White and Johnson, 1980).

No current specialists may be available for some taxa in urgent need of thorough, monographic work. These taxa would be identified by or highlighted in the kinds of publications listed above. Research on such taxa might even be financed by a national biological survey.

A national biological survey functioning in the role of research coordinator and general sounding board may lead to the discovery of needs for a variety of other kinds of publications. A need may be found to convey scientific concepts to laymen and to bring together technical information important to them in their work. If extensive, monographic information on *Cannabis*, for example, were not already available (Small, 1979), a national biological survey might have been the impetus to organize a task force to conduct such studies and may then have provided the editorial team necessary to see the resultant volumes through to completion.

Examples of Projects

Specific needs for a comprehensive survey of the insect fauna of America have been identified by the entomological community in North America. Plans for an Insect Fauna of North America project have been under consideration for some years now by the Entomological Society of America. Having identified the need, the first priority was to determine the scope of this task—how many taxa would be treated by this project and how well each was known. The second priority was to inventory the available human and literature resources (Arnett, 1983).

The Flora of North America project is another good example of the steps a national biological survey might take toward producing a publication. The need for a publication on the plants of the U.S. and Canada has been recognized for many decades. *North America is the only major temperate area without such a publication.* Information on North American plants is dispersed in monographs and articles and in regional and local publications. No unified synthesis of the treatments of different floras has been published and many groups have never been studied in detail. In 1965, the North American botanical community began work on a project to produce a flora. That project continued until 1971, when lack of funding forced suspension. The need has remained, however, and in 1982, 20 North American botanists met to discuss the possibility of trying again to establish such a project. In 1982, the Canadian Botanical Association reaffirmed its support of such an effort, and in 1983, the American Society of Plant Taxonomists passed a similar resolution. The need has been recognized and defined: we need a synoptical flora that will cover all of the vascular plants north of Mexico. We need more than that, but right now that is what is feasible for the botanical community to produce.

Once the need was defined, a core group of plant taxonomists began the preliminary work necessary to progress to the writing of the flora; this preliminary work is still underway. We are assessing the human resources and the literature to determine our level of knowledge for each taxon, which of the many groups are already being studied by specialists, and for which groups specialists are lacking. Having this information will allow us to alert the community to the immediate needs of the project and to plan for expertise that will have to be provided by the project staff. An editorial committee has been organized, and plans for soliciting and reviewing manuscripts have been made. A databank will be an integral part of the project and will allow publication of up-to-date atlases, lists of chromosome numbers, etc., as well as provide a wealth of other kinds of information about the flora. The flora project will synthesize the findings of recent work and stimulate further research on the plants of North America. The multi-volume work will be a compendium of information on the North American flora and a practical aid to identification. The steps taken in this project, which would also be the logical steps to be taken by a national biological survey, are as follows:

1. Identify the need for a publication,
2. Assess the resources to determine what expertise the community could provide and what must be provided by project staff,
3. Identify the potential users of the project and plan to design the publication to be useful to as many users as possible,

4. Design a workable organization for the project and proceed toward the writing of the publication.

In the case of the insect fauna of North America, substantial logistical and financial support may be needed from a national biological survey because of the enormity of the task and the current lack of experts in critical groups. In the case of the Flora of North America, the botanical community is already making fair progress on the organization of this project: the experts and necessary staff and editorial expertise are available, the organizational structure is in place. The publications could quite logically be considered part of a series produced under a national biological survey umbrella.

RECOMMENDATIONS

The steps that a national biological survey should take relative to producing publications are:

1. Inventory Resources,
2. Assess Needs,
3. Coordinate Efforts,
4. Sponsor or Facilitate Publication,
5. Disseminate Information.

LITERATURE CITED

Arnett, R. H., Jr. 1983. *Status of the taxonomy of the insects of America north of Mexico: a preliminary report prepared for the Subcommittee for the Insect Fauna of North America Project.* Entomological Society of America. College Park, Maryland.

Berggren, G. 1903. *Atlas of seeds and small fruits of Northwest-European plant species (Sweden, Norway, Denmark, East Fennoscandia, and Iceland) with morphological descriptions.* Berlings, Arlov.

Boomsma, C. D. 1981. *Native trees of South Australia.* Woods and Forests Department, South Australia, Bulletin 19, Second Edition.

Cerceau-Larrival, M.-T. 1980. Umbelliferae, Hydrocotyloideae, Hydrocotyleae *In*: Nilsson, S. (ed.). 1980. *World pollen and spore flora 9.* The Almqvist and Wiksell Periodical Company. Stockholm.

Codd, L. E., B. De Winter, & H. B. Rycroft. (eds.). 1966-. *Flora of Southern Africa.* Cape and Transvaal Printers Limited, Republic of South Africa.

Dyer, R. A. 1975. *The genera of southern African flowering plants.* 2 Volumes. Botanical Research Institute. Dept. of Agric. Tech. Services, Pretoria.

George, A. S. 1981. *Flora of Australia.* Bureau of Flora and Fauna, Canberra. Australian Government Publishing Service, Canberra.

Holmgren, P. K., W. Keuken, & E. K. Schofield, (compilers). 1981. Index Herbariorum Part I *The herbaria of the world.* 7th edition. Dr. W. Junk B. V., The Hague/Boston.

Jalas, J. & J. Suominen (eds.). 1983. *Atlas Florae Europaea: distribution of vascular plants in Europe.* The Committee for Mapping the Flora of Europe and Societas Biologica Fennica Vanamo.

Jessop, J. (editor-in-chief). 1981. *Flora of Central Australia.* The Australian Systematic Botany Society. A. H. and A. W. Reed Pty. Ltd. Sydney.

Jessop, J. P. (ed.). 1983. *A list of the vascular plants of South Australia.* Adelaide Botanic Gardens and State Herbarium, and the Environmental Survey Branch, Department of Environment and Planning, Adelaide.

Kiger, R. W., T. D. Jacobsen & R. M. Lilly. 1981. *International register of specialists and current research in plant systematics.* Hunt Institute for Botanical Documentation, Carnegie-Mellon University. Pittsburgh.

Komarov, R. L., et al. 1934-64. *Flora S.S.S.R.* Moscow/Leningrad. AN S.S.S.R. Press.

Lovett, B. H. 1970. *The geographical distribution of native plants in South Australia and the Northern Territory: an index to articles in selected South Australian periodicals.* Volume 1. Research Service Bibliographies Series 4, No. 136. State Library of South Australia. Adelaide.

Moore, D. M. 1982. *Flora Europaea check-list and chromosome index.* Cambridge University Press, Cambridge.

Small, E. 1979. The species problem. *In*: Cannabis: *science and semantics.* 2 Volumes. Corpus, Toronto.

Tutin, T. G., et al. 1976. *Flora Europaea.* Cambridge University Press, Cambridge.

White, D. J. & K. L. Johnson. 1980. The rare vascular plants of Manitoba. *Sylogeus* 27.

Applications and Use of Biological Survey Data

Robert E. Jenkins, Jr.
The Nature Conservancy

Abstract: The concept of a national biological survey is endorsed and the main applications of such a survey are enumerated. A general framework for the scope, structure, and operations of a national biological survey are suggested. An existing system, the network of State Natural Heritage Inventory Programs, is reviewed as an incomplete but advanced model and a possible foundation for expansion for a national biological survey. Some currently unmet needs are listed that might be dealt with by a national biological survey through research or intensification of inventory. Several recommendations are made about how to proceed in further encouraging the establishment of a national biological survey.
Keywords: Biological Survey, Biological Diversity, Inventory, Conservation, Natural Heritage.

INTRODUCTION

We need a national biological survey. For that matter, we need a national ecological service. We need a major agency or institution that is interested in our biota and ecosystems for reasons that transcend immediate economic returns or consumptive use. It is very shortsighted of the U.S. not to recognize the importance and practicality of knowing everything we can about our biota and ecosystems. Many other nations have done a much better job in these areas than we have. When you stop and think about it, it seems little short of incredible that at this late date our biota are still taken so much for granted instead of being universally recognized as the very basis for ecosystem health and of every renewable natural resource on which we might someday want to make use.

Perhaps with all the media attention given lately to conservation of biological diversity, genetic resources, etc., it will finally dawn on people that it is unwise, improvident, and probably dangerous for us to blunder on in bliss of ignorance about our biota. If the current groundswell denotes any real commitment, we should seize on the opportunity to initiate a vigorous program to complete the basic catalog of our national biotic heritage and accelerate our efforts to discover much more about it.

In this paper, I hope to further these efforts by suggesting what I think should comprise a national biological survey, describing an extant partial model, discussing some applications, pointing out some lessons learned, and listing some currently unmet needs that a national biological survey might incorporate in its programs.

UTILITY OF A NATIONAL BIOLOGICAL SURVEY

We must make clear the practical applications of a national biological survey, of which there would seem to me to be several:

1. *Conservation planning.* Everyone seems to be beginning to understand that our biota and gene pools are vital resources that we must take measures to conserve. Next, everyone might be made to understand that we cannot identify and design biological reserves effectively unless we know what our biota are, where they are to be found, and what they require to live.

2. *Environmental impact assessment/environmental quality regulation.* Biological aspects of environmental impact analysis and development planning could be much enhanced by the availability of more and better organized information such as could be produced by a well organized survey. To some extent this is the other side of the coin from direct conservation planning, and it is just as important in averting unnecessary destruction.

3. *Cataloging genetic resources.* Every projection seems to be for the rapid evolution of U.S. industry toward dominance by the high technology fields, and one of the most important of these is supposed to be genetic engineering. The universities appear to be hell-bent on transforming their biology departments into gene-splicing companies, but as yet, precious few are showing any concern for the study or conservation of the gene pools upon which such endeavors depend.

4. *Land, ecosystem, species, and natural resource management.* At this point we really know very little about the habitat and ecological requirements of individual species, much less of species assemblages or entire ecosystems, nor do we have more than a glimmer of an idea about how they respond to our current land management practices. Gaining knowledge in these areas is not only pertinent to managing nature preserves, but such knowledge could undoubtedly be important in "multiple use" situations. A lot of what we currently do in agriculture, range management, etc., involves the use of unadapted exotic species and artificial systems that are probably not performing to the level of which native species and systems would be capable, even in productivity and economic return, much less in terms of ecosystem stability and function.

5. *Facilitating biological and ecological research.* Expanding our knowledge of our native biota and organizing that knowledge base in a more effective way than it currently is organized in our varied and scattered repositories would greatly assist us in learning even more. Better data organization would almost certainly facilitate the advancement of scientific theory as well as the further accumulation of factual information.

STRUCTURE OF A NATIONAL BIOLOGICAL SURVEY

Unproductive confusion will arise if we continue to discuss a national biological survey without defining what we think it should be. It is evident that among the symposium participants there are some fairly different ideas about the structure and functions of a national biological survey. The various presentations here contain a $100 million worth of respectable ideas, but somebody will have to decide how to get and spend the first million—on that will depend whether a national biological survey dies aborning. Achieving success in the longer run will be an even harder matter. In any case I can't say what I think about a national biological survey unless I first specify what I think it would be. I will try to incorporate some ideas of priorities for the expenditure of the first million.

First, I would like to see an integrated national institution, not just some sort of loose consortium. I rather liked Orie Louck's diagram with its central coordinating unit. A variety of existing institutions such as museums, universities, agencies, and organizations should definitely play important parts. Many on-going efforts deserve continuation and enlarged support in this connection, but we really need a unified central core.

Second, there seem to me to be three main operational needs:

1. *An information repository with information synthesis functions.* As Allan Hirsch points out, a massive amount of biological information collection is going on, but as Steve Edwards has often observed, it tends to be a "one-time use, throw-away item." We need to do a vastly better job of capturing, storing, and organizing such information and in ways that facilitate practical use. I agree with Nancy Morin that we will need to produce a variety of published reports from this information, but we definitely need to generate these from easily updatable, continuously maintained, computer databases.

2. *A methodological development unit.* A great many ambitious programs fail because insufficient importance is attached to methodological development and refinement. Particularly in an undertaking such as a national biological survey, where standardization among so many parties must be achieved, truly superior methods are critical. These methods can only be developed through an intensive, continuous effort. I consider this to be an absolutely top priority undertaking for the central staff.

3. *A mechanism for research and data gathering, principally on a widely distributed, multi-institutional (but coordinated) basis.* This needs to be done on an interative basis, with limited resources allocated to a succession of reasonably discrete tasks that can produce products or services of direct use. An early task might be to survey other existing data collection efforts with a cold eye, to produce a sort of data bank of data banks (the term "data bank" is used to denote an assemblage of manual as well as computerized data) that would clearly identify the relatively few such data banks that are truly valuable.

SCOPE OF A NATIONAL BIOLOGICAL SURVEY

There is probably less difference of opinion on the scope of a national biological survey than on its structure, but there is likely to be considerable disagreement about relative priorities. The obvious subject matters that should be of interest include:

Biological taxonomy. The national community of systematists is one of the major forces pushing for a national biological survey at this time, as a vehicle for rapidly advancing their collection activities and research to at least identify and classify all the components of our biota. I gather that the entomologists have a long way to go to even identify and name most of their species; other groups clearly need major work as well. Beyond taxonomic monographs, it would be useful to produce checklists and diagnostic keys and to characterize species and infraspecies in a variety of ways and for various purposes.

Biogeography. Most of us would agree that we also need to obtain a much clearer picture of our biota's distribution.

Species autecology and population biology. Most will probably agree that part of the business of a national biological survey should be to find out about species characteristics, ecological requirements, and functions.

Synecology. Classification of biological communities and research on ecosystem functions are logical extensions of the above lines of inquiry, but they would add much additional complexity and expense to the work of a national biological survey. I liked Orie Louck's figures on biome data, and they seem to fit in with Allan Hirsch's prescription for some EPA user needs, but both are assuming that a national biological survey will be heavily oriented toward system ecology, and this is probably not the way to spend the first million.

Genetic research. Most of us believe that we need to know a great deal more about genetic diversity, its distribution through species and populations, mechanisms of inheritance, problems of inbreeding and outbreeding depression, genetic aspects of minimum viable population sizes, and the importance of gene pools in human affairs. Again, this would become much more involved and could even lead into the entire "conservation biology" agenda. Many of us consider these lines of inquiry very important, but until a national biological survey has taken on a more definite character, we cannot be sure that it is the best vehicle for their advancement.

Geographic scope. It is obvious that a national biological survey would cover all 50 states, and for jurisdictional reasons, most people would probably agree that it should cover all U.S. territories and possessions as well. Some will argue that it would be eminently reasonable to extend it to the rest of North America through some form of cooperative venture with Canada and Mexico. A major point of debate is apt to be whether a national biological survey should have an international component beyond North America, and if so, what form it should take.

Lastly, it occurs to me that there may be a need for some educational initiatives to accomplish or facilitate various aspects of a national biological survey. We may not, for example, have enough specialists in certain fields to deal adequately with some taxonomic groups, and it may be necessary to take measures to provide additional manpower.

STATE NATURAL HERITAGE INVENTORY NETWORK: AN EXISTING BIOLOGICAL SURVEY MODEL

I addressed the Association of Systematic Collectons annual meeting on the subject of State Natural Heritage inventories several years ago when it met at Harvard. In 1983, the organization was kind enough to give the Conservancy its annual Award of Excellence for this work, so for many ASC members this will just be something of an update. I will describe the network of these programs as constituting something of an existing national biological survey and discuss some of the uses made of the data, some of the limitations in its scope, and what else we would like to see developed to fill currently unmet needs. I will be addressing these matters from the perspective of The Nature Conservancy, which functions in several relevant capacities—as a developer of data collection and management methods, a compiler and organizer of existing information, a gatherer of new information, a data user, and a supplier of information for the uses of others. In these roles, we have been involved to some extent in all five of the applications I've suggested as appropriate to a national biological survey (but very little so far in genetic resource analysis at a genotypic level).

The Nature Conservancy's entire program is directed to the preservation of biological and ecological diversity in this country, and, increasingly, elsewhere in the world. The Conservancy's precursor, a special conservation committee of the Ecological Society of America, was established in 1917. The early activities of that committee were concentrated on the identification through surveys (or "inventories") of biologically and ecologically significant "natural areas" for preservation. Along with the direct preservation and long term management of such areas, this activity continues to the present day. In the field of biological conservation, no other institution has as much experience in inventory, information management, or application of information to decision-making as does the Conservancy.

Much of the early inventory activity was of only limited value for reasons that need to be understood by anyone else contemplating similar biological surveys. The reasons have been explained at greater length elsewhere (Jenkins, 1982), but in summary, the main shortcomings were imprecise goals, discontinuity of effort, overambitious scope for available resources, lack of standardized methods and terminology, and general failure to patiently organize a disciplined effort. These problems began to be solved in 1974 when, along with the South Carolina Wildlife and Marine Resources Department, we launched the first "State Natural Heritage Inventory" program.

State natural heritage programs are "permanent and dynamic atlases and databases on the existence, characteristics, numbers, condition, status, location, and distribution" of biological species, natural communities, and other entities, features, or phenomena collectively referred to as "elements of natural diversity." Unlike previous inventories, natural heritage inventories have none of the fatal shortcomings listed in the last paragraph (although they could certainly benefit from increased funding to meet their admittedly ambitious goals).

In 39 other states, as in South Carolina, state natural heritage inventories have begun as cooperative ventures between state governments and the Conservancy, to be carried on in the long term as units within state agencies. Cooperative programs have also been undertaken with the Tennessee Valley Authority and the Navajo nation. In most of the other states, such inventories are currently being carried out by the Conservancy alone, by the state alone, or by a multi-institutional arrangement. Thus these inventories are now collecting and organizing biological information across virtually the entire nation. Collectively these programs have annual operating budgets of over $7 million, employ over 200 biologists and information managers, and call on the knowledge and voluntary assistance of thousands of scientists and amateur naturalists. One measure of the extent of this effort is that these programs collectively manage information on over 150,000 individual rare species populations and localities. Incidentally, the

Conservancy's international arm is helping to set up quite a number of heritage inventories in various Latin American countries, where they are usually referred to as "Conservation Data Centers."

For a number of reasons, the state natural heritage programs do not quite constitute a complete national biological survey, at least with the dimensions I proposed at the beginning of this paper, although they are generally evolving in that direction.

The main reason that the heritage programs do not constitute a complete national biological survey is that their funding is too limited. For one thing, they cannot afford to gather and organize information on all species, so they concentrate most of their efforts on rare and endangered species because these are the ones that require conservation attention. Even here they primarily focus on vertebrate animals and vascular plants, although they do have information on a small fraction of the invertebrates and non-vascular plants that specialists believe to be very endangered, with rather better coverage of such popular groups as freshwater molluscs and Lepidoptera. Even on rare vertebrates and higher plants, they can only afford to transcribe and computerize a fraction of the information on species biology, potential utility, etc., that the Environmental Protection Agency or the U.S. Department of Agriculture would consider ideal for some of their purposes.

From our experience in gathering and managing information on the better known and rarer taxa, it is obvious that to expand coverage to the rest of the groups will only be possible with annual budgets of tens of millions of dollars. This would be true even if the taxonomy of these taxa were already fairly well known and we were satisfied with much more limited information on many subjects than we currently collect on rarities. For example, we might collect and record distribution data on abundant taxa to county-of-occurrence precision, but no one could seriously contemplate collecting data on every known stand of white oaks, as we have for hundreds of rare plant species. Even if we suppose that much of the information and the infrastructure for supplying it already exists in the universities and museums, experience suggests that just the central information managing unit itself will need an annual budget of several million dollars, and funding the central unit must be the first priority.

Heritage programs attempt to make up for their limited taxonomic coverage by expending up to half their energies on classification and inventory of ecological communities on the theory that by identifying and preserving good examples of major terrestrial and aquatic community types, we will provide habitat for as many as possible of the species about which we cannot afford to collect individual information. This focus and emphasis placed on rare and endangered species are efficiency measures, and we originally imagined that this approach would save us the trouble of dealing with up to ninety percent of even those vertebrates and vascular plants that are not in extreme peril. Ironically, however, since everything is rare and endangered somewhere in its range, we eventually realized that one or another state program was collecting information on nearly every native species in these groups.

Moreover, even the rare species are usually found in more than one state, so we gradually began to see advantages to centrally storing a growing body of information at our national office to increase the efficiency of collection and data use by several states. Since at least some basic data existed somewhere (wherever the species is rare) in the network for all vertebrates and higher plants, the national databases began to be more and more comprehensive for these groups, at least at a limited level of detail.

These central databases have brought together an array of biological information never before assembled for the entire U.S. For example, the national databases contain the most complete computerized checklists in existence for all U.S. vertebrates (from many sources) and vascular plants (from John Kartesz' latest unpublished synthesis). Manual files have been assembled on a large number of these species, including all of the rarer and more endangered ones. We have computerized "Element Global Tracking" records on 23,927 North American species or varieties of plants (22,546 full species) and on 5,952 vertebrates (3,703 full species including marine fishes and sea birds). In these records, all of the vertebrates and about half of the plants have been assigned range-wide endangerment rankings on the basis of standardized factorial analyses. In many instances, the specific "Element Global Ranking" records themselves are computerized and the rest are on manual forms. Included is such information as numbers of populations, number of individuals, etc., and these databases are being expanded every day. We are also computerizing "Element State Tracking" records, which should shortly give us a complete state-of-occurrence picture of known vertebrate distribution for the U.S. At the current rate of input, however, it will be at least several years before we will have achieved the same for vascular plants. Because there is a great demand for information on vertebrates, we are currently expanding this part of our national and state databases by adding 30 or 40 data fields on each species.

The Nature Conservancy has a special heritage task force in its national office charged with maintaining this growing volume of centralized information, as well as for developing and refining methodology, documenting standard procedures, and training and supervising new state level staff. A basic rule for division of data responsibility within the cooperating network is that information on individual populations, localities, and protected areas is stored at the state level, whereas information on generally applicable species biology eventually flows upward to the national data bank. Additional information is managed centrally to help maintain standardization on such matters as taxonomic classification so as to facilitate the exchange of information, reduce confusion, and enable us to bring data together on a regional or national basis for such things as range-wide status analyses. This central data bank is contributing materially to increased efficiency by reducing duplication of effort.

Lead responsibility for maintaining an overall master file on at least each individual rare species is gradually being assumed by individual state heritage staffs, usually on obvious distributional grounds. Responsibility for other species, such as the intrusive exotics, is being taken by Nature Conservancy stewardship staff or outside cooperators. From these sources, the central databases are being constantly refined. The updated information is shared throughout the network with all who make a reasonable request. For the rarest species, increasing numbers of individual populations are being monitored annually. Some of this information is being added to the central databases as well. In addition to the state heritage programs, the national staff, and the stewards in the regions and state offices, heavy input to the central databases is coming from special units of the national task force located in the Conservancy's four regions.

USES OF THE DATA

One of the main uses of biological information from the heritage network, and the one for which it was originally intended, is for systematically planning for *in situ* conservation of biological diversity, or in other words, to identify and set priorities among areas that should be preserved for this purpose. This is done primarily at the state level by generating "natural diversity scorecards." These consist of species and community lists arranged in order of decreasing relative endangerment, along with their existing occurrences on the landscape, a ranking of these on the basis of quality and viability, and the further translation of this information into specific priority sites that are for designing nature preserves. In many states where heritage inventories are more mature, this has transformed biological land conservation as practiced by the Conservancy and a growing number of other cooperating agencies into an objective, data-driven process. From the point of view of endangered species, most of which are endangered in large part by habitat destruction, this has become the most pervasive mechanism by which they are being saved from extinction or further depletion in this country.

A second major use of the heritage data, and one which is of the utmost importance to state governments, is to provide yardsticks for environmental impact assessment and the review of potential development projects. The existence of comparative information on so many species, communities, and localities allows us to get at questions such as that posed by Allan Hirsch about the relative significance of damage to a habitat or species population. The comprehensive series of U.S. Geological Survey topographic quad maps, which every program uses to plot actual localities of endangered species occurrences and outstanding community remnants, provides an efficient means of identifying critical resources that could be affected by potential developments. This is particularly important in regard to decisions about which of two or more potential sites will be developed for any given purpose. In cases where the proposed development cannot be relocated away from a vulnerable site, additional heritage information about biological tolerances can be employed to recommend modifications of the project design to mitigate environmental impacts. Prior to the time this sort of very specific locality information became available, project reviewers had been forced to rely on inferential guesswork. Many of the state heritage programs participate in hundreds or even thousands of project reviews per year. In addition to averting the destruction of important biological entities and areas, the early and definitive application of this information helps to avoid a lot of unnecessary resource expenditures and conflicts.

Another important use of heritage information is in species status review to ascertain and document degree of endangerment. In some states (and lately, on an aggregate multi-state basis), scores of species that were candidates for endangered species listing on the basis of more or less anecdotal information have been dismissed from further consideration after newly collected, compiled, or aggregated information showed them to be much more widespread and abundant than previously believed. Other species have been shown to be rarer than once thought.

These sorts of results show how important good biological information can be in cost-effectively allocating scarce resources. To achieve this result, however, heritage inventories employ concepts of incremental data accumulation and iteration to constantly refocus information collection efforts on the next most important thing to learn. This keeps them from becoming bogged down in impossible volumes of less important data so that, for example, they can get around to such tasks as *de novo* searches for new occurrences of inadequately known species. An important lesson or two is here for whoever is to undertake an even more comprehensive biological survey where being buried in data will be an even larger problem.

The same priority-setting mechanisms are allowing heritage programs to focus on questions of importance to ecosystem management, at least in regard to natural areas. For examples, on the one end of the species spectrum, heritage databases focus our attention on those species that not only require habitat protection, but also management intervention. To intervene intelligently, we need more information about the biology of these species, their life history strategies, ecological tolerances, etc. On the other end of the spectrum, there are some very abundant species, such as intrusive exotics, about which natural area managers need to know more to effectively control them. In a more comprehensive biological survey, this sort of thing would probably be greatly expanded because of special interests in the biology of economically important game species, crop relatives, disease vectors, timber trees, etc. The structure of these databases brings such data needs to the top of our priorities in the order they really merit, so that we can efficiently focus our research efforts or persuade others to do so.

Nearly all of these applications require information at both national and state levels. We have found that it is simply not practical to maintain current, accurate information on individual localities or populations at the national level, nor is there much direct use for such information at that level. Therefore, some sort of network involving operating units at the state level is absolutely crucial to meet the most pressing decision-making needs.

UNMET NEEDS UPON WHICH A NATIONAL BIOLOGICAL SURVEY MIGHT FOCUS

Hirsch, Klassen, and others have treated a number of needs within their agencies and fields, so I am confining myself mainly to problems in the field of biological conservation:

1. *Species taxonomy.* Others can speak more authoritatively on needs in this area, but standardized taxonomy is very important to us in our survey and conservation work. A standardized North American Flora would be extremely helpful to us, and although separate initiatives are being taken on this right now, it is a logical part of a comprehensive biological survey.

2. *Additional leads to endangered species of invertebrates and non-vascular plants.* We have incorporated into our existing surveys all species in these groups that specialists have recommended, but there is an awkward chicken-and-egg aspect to this in that the specialists need to know more about the species before they can decide whether to consider them endangered, but we do not collect information on these species until somebody says they are endangered. The level

of background knowledge simply needs to be elevated by whatever can be accomplished through a national biological survey to expand research on these groups. I think we need to be realistic about this, however. As Barry Chernoff showed, there is probably a point at which cricket taxonomy, distribution, etc., changes faster than we can hope to keep up. As far as conservation goes, we are most apt to save populations of such things by preserving examples of ecosystems and communities.

3. *Support for more field surveys.* Even with an information management system that makes priority setting very systematic, heritage staffs are hard pressed to conduct all of the field surveys that they need to, even just to verify earlier records of species and community occurrences garnered from literature, museum collections, etc. On top of that, a lot of *de novo* searching is required. We get substantial help in the field from university scientists and other volunteers, but for endangered species, time is of the essence. Increased funding for field surveys could be a life-and-death matter for some species.

4. *Support for applied research on selected species (and communities).* In many instances very little is really known about the biology and ecology of a given species, and if it is not prospering we frequently haven't a clue about how to intervene in its management. Again, much of the needed research is only going to get done in time for us to use it if a source of increased funding is identified.

5. *Genetic research.* There are some very pressing genetic questions of the utmost importance to conservation decision-making that are not now being answered. If these lines of inquiry are to be pursued as part of a national biological survey, I would be particularly interested in research that could help clarify just how important it is for us to preserve subspecies, ecotypes, or peripheral and disjunct populations. Do these taxa characteristically have unique alleles or irreproducible genetic complexes? Would their loss constitute the loss of a thousand years of selection for local adaptations, or do these populations merely represent trivial recent or recurrent invasions of marginal areas? At present, it seems prudent to assume that it is important to preserve such local populations so that a large part of our conservation resources is spent on state rare species, etc.

6. *Reliable longterm funding.* I have mentioned several aspects that we would make haste to accomplish if we had a source of increased annual funding, but from my perspective, the biggest single shortcoming of the natural heritage network is the lack of reliable sources of permanent funding that could guarantee that the cumulative body of knowledge and experience will not someday be undone. Therefore I agree with Stan Shetler that reliable long term funding is everything. The prospect of greatest interest to me is that a national biological survey be established as an institution with such a firm financial basis that the central information management operations could be secure for the long term.

7. *Other.* There are many additional things that the users need, even just in my own field of conservation. If the national biological survey included sufficient funding for a substantial research effort beyond the more basic aspects mentioned above, I would like to see it take on the whole research agenda of what some persons are calling "conservation biology," which includes questions about biological preserve size, other aspects of preserve design, minimum viable population

sizes, effects of ecosystem fragmentation, etc. I also would like to see much more done, on a longer term and a more disciplined basis, on ecosystem research and monitoring. This research could be partly basic, such as most of the research being sponsored by the National Science Foundation through its Long Term Ecological Research program, and partly applied, such as that involved in management of grassland ecosystems by use of fire. It would be very useful if some special unit of the national biological survey were able to devote itself to the extremely complicated and difficult questions of community and ecosystem classification, as Alan Hirsch and Christine Schonewald-Cox have suggested, since this has been such an intractable obstacle to various aspects of conservation and system management.

CONCLUSION

For several reasons, I have devoted much of this paper to discussing the State Natural Heritage programs at some length, rather than taking a more disinterested perspective. First, they are what I know the most about, and if I am to give advice I should do so on subjects where my judgement is least apt to be mistaken. Second, it seems to me that the network of natural heritage programs is not sufficiently known among scientists advocating a national biological survey, or not sufficiently appreciated, because until now it has not figured as weightily in discussions as the facts would seem to justify. At this time, these programs collectively constitute the closest thing to a national biological survey that exists and as such should serve as a relevant model to be examined, a possible foundation upon which to build, or at least a contributing part of the grander survey we all hope will ensue.

As for exactly where the heritage data network fits into a more comprehensive effort, we feel quite flexible and open to suggestions. The Nature Conservancy intends to fully cooperate with any intelligent plan, even one in which the heritage operations would be wholly absorbed. Just as we have transferred full control of individual state heritage programs to the state governments, which are better able to afford their long term support, we have always been prepared to hand over our national task force functions to a more muscular long term sponsor. However, we began to do this once before with the late Heritage Conservaton and Recreation Service, and if we are to try again, we would like to be sure of the viability and soundness of the new management unit.

This said, I sincerely hope that the current initiative will lead to the establishment of an institution capable of a much larger program of biological survey and information management, including the full range of associated genetic and ecological research work. However, the reasons why such an institution has not already been established still persist as impediments to the current effort. If such a thing is to be accomplished, it will take a much more cohesive effort from the community of biologists, speaking with a more united voice, than we are used to seeing when such matters are brought to public attention. In doing this, it will be important for us to make one thing clear—that although biologists are calling for a biological survey, this is not just special pleading. Most of us are not calling for more biological knowledge because we are in the business; instead, we entered the business because we saw the need for more biological knowledge.

RECOMMENDATIONS

1. Continue to explore the complex issues involved in creating a national biological survey in a disciplined manner. This could involve establishing an advisory or planning committee, finding an independent body to study designated facets under contract, and holding a series of workshops to involve a wide array of interested parties.
2. Specific issues to be studied include the following:
   a. Optimal structure and scope of a national biological survey.
   b. Review of current activities and institutions and their possible contributions.
   c. Examination of national biological surveys in other countries as possible models.
   d. Previous attempts to establish new programs or institutions of a related nature, both successful (e.g., National Center for Atmospheric Research) and unsuccessful (e.g., National Institute of Ecology), to identify positives and negatives that might be applicable to the current undertaking.
3. It may also be useful to explore the attitudes and opinions of a wide spectrum of individuals in Congress, federal and state government agencies, the nonprofit sector, and possibly in business to discover probable support or resistance to the idea of a biological survey, and specific suggestions for its establishment.
4. Several related initiatives being considered at this time are receiving high level attention, mainly concerning the need for greatly expanded efforts in the conservation of domestic and international biological diversity. The relationship of these matters to a possible national biological survey should be explored to avoid duplication or collision between these ideas and the actions that might result.
5. Considerable thought should be devoted to questions of budget levels and sources of funding. There would be little value in attempting to conduct a national biological survey on a shoestring budget, and since a biological survey needs to be a permanent activity and not just a short term project, reliable annual funding is crucial. If I have not explicitly said so yet, this means to me that realistically the national biological survey must be a government institution or one with permanent government funding.
6. The sections of this paper on structure, scope, functions, and current users' needs constitute additional specific recommendations in themselves. In particular, I think that the existing State Natural Heritage network should be explored as a model, a possible foundation, or at least a contributing part of a more complete national biological survey.

LITERATURE CITED

Jenkins, R. E., Jr. 1982. *Planning and developing natural heritage protection programs.* Presented at the Indo-U.S. Workshop on the Conservation and Management of Biological Diversity in Bangalore, India. Copies available from The Nature Conservancy.

Jenkins, R. E., Jr. 1984. An example from America. *Naturopa* 48: 10–11.

The Nature Conservancy. Unpublished. *Natural heritage program operations manual.* About 400 p. Not distributed, but available for inspection at Nature Conservancy and state natural heritage offices.

ADP Technological Perspectives of Biological Survey Systems

H. Edward Kennedy
Maureen C. Kelly
BioSciences Information Service (BIOSIS)

Abstract: Given the characteristics and magnitude of a national biological survey, it is essential that provision be made for managing the large volume of information that has been and will be collected. The value of the survey will depend in part on the availability and usefulness of the information collected. This paper reviews some of the short and long term technological developments that might impact on the capture, storage, manipulation, and use of biological survey data. Recommendations are given for the development of a "referral data base" as the mechanism for coordinating the data collection and data management aspects of the survey. Also recommended is the establishment of a central agency for the registration of organism names.
Keywords: Information-handling, Data Management, Technology, Referral Data Base, Optical Disks, CD-ROM Disks, Microcomputers, Telecommunications, Networks.

INTRODUCTION

This paper reviews some of the short and long term technological developments that might impact on the capture, storage, manipulation, and use of biological survey data. The review was conducted from the perspective of a user of data processing hardware and software. Our employer, BIOSIS, is a not-for-profit company, which abstracts and indexes biological information and makes it available to users in printed and computer-readable form. Over the past 30 years, computers have become a critical component of our production operations.

Because we represent a user of technology, we are familiar with the problems as well as the advantages of computerization. We have learned to take with a grain of salt the glowing promises of advertisers. For these reasons, we have taken a somewhat conservative approach in preparing this paper. After all, we would not want to be guilty of perpetuating any of the fallacies Dr. Shetler so clearly identified for us in his 1974 paper on "Demythologizing Biological Data Banking." Still, we hope that you find new opportunities for information management in some of the technological advances we will describe.

The task of conducting a national biological survey involves many practical problems in information handling. Given the characteristics and magnitude of a national survey, it is essential that provision be made for managing the large volume of information that has been and will be collected. The value of such a survey will depend, in part, on the availability and usefulness of the information collected.

Over the past 30 years, advances in computer technology have had a significant impact on the way in which we capture, store, manipulate, and distribute biological information. Few would question that these changes will continue, most likely at an accelerated rate. And because the use of new technology inevitably lags behind its development, many, if not most, of the changes we will see in the next five years will result from technology already developed or under development.

We have attempted to pull together highlights of the technology, both existing and under development, which, in our view, could facilitate the management of information collected by a national biological survey. Of necessity, much will be left unsaid or will be dealt with in a cursory manner. For convenience, we have chosen to organize the technology into four categories—technology that facilitates:

I. Information Capture
II. Information Storage
III. Information Processing
IV. Information Use

(Be assured that there is nothing hard and fast about these four categories; many technological developments could easily be discussed in more than one category.)

I. INFORMATION CAPTURE

In the area of information capture, significant technological advances have been made in recent years. The 80-column punched card, so common in the past, is becoming an endangered species today.

Data capture for a national biological survey brings with it a variety of requirements. Of particular interest are technologies that provide for direct capture of data in the field as well as those that enable existing, printed data to be incorporated into the new data collection.

In the past, the closest one could get to direct capture of data in the field was to have the researcher encode his findings on "mark-sense" forms that would later be optically scanned. This remains a cost-effective approach to collecting encoded data, and it is being used in France to collect data on the distribution of French flora and fauna.

However, a great many more options now exist for capturing data at its source, largely as a result of microcomputer technology. In just the last five years, micros have come down in price and size to the point where it is now feasible to use them in the field. Handheld computers, derived from the high-end, alpha-numeric calculators, are the fastest growing segment of the portable computer industry. Advances in flat-screen technology and improvements in battery power sources are expected to yield significant improvements in briefcase-size computers.

Equipped with such specialized user interfaces as a voice translator, light pen, scratch pad, a "mouse", or a touch-sensitive screen, microcomputers are becoming increasingly user-friendly. Technology that supports voice input and output is so practical that limited versions can even be found in toys. Computers are now available that can synthesize speech and/or respond to verbal instruction.

Data captured on a micro can be stored on non-volatile media such as floppy disks or transmitted directly to a host computer over a telephone or via radio signals. It is expected that before 1990 a commercial system will be available that will let people communicate with distant computers via satellite by use of a small pocket transceiver.

The capture and integration of existing data presents a different set of problems. Much of the existing data is available in printed form only, either as books, catalog cards, or field notes. Advances in optical-character-recognition technology now make it possible to scan material printed in a wide range of type fonts and convert it into digital form. In particular, the Kurzweil scanner can recognize thousands of combinations of typefaces and type sizes in a wide variety of page formats. It digitizes documents up to 25 times faster than keyboard operators and accomplishes its conversions with surprising accuracy and flexibility.

In some cases, it may be necessary to create a digital record of an "image," as in the case of pictures, drawings, and occasionally text. New technology is being developed in this area also. Companies such as IBM, Wang, and Microtek are beginning to market products that can scan and digitize anything on a piece of paper. Once digitized, the image is stored on a microcomputer disk from which it can be manipulated by use of special software. In combination with telecommunications equipment, the image can even be transmitted to other locations in the same fashion as facsimile. Although this technology provides many useful opportunities for capturing graphic material, it should be understood that storage requirements for images are high. An 8-1/2 x 11-inch image requires about 500 thousand bytes (500K) of disk storage (when digitized at 200 pixels per inch).

II. INFORMATION STORAGE

Information storage technology is probably the area of greatest interest to those responsible for managing the information generated by a national biological survey. Developments in storage technology over the past 20 years have been dramatic. The key areas where major developments are now taking place are: magnetic storage, optical disks, and combination technologies.

Magnetic Storage

In magnetic media, data are recorded by repeated polarization of tiny areas along the surface of the medium. The orientation of the poles is used to indicate a "0" or "1." Storage capacity is affected by the size of the area polarized and by how closely together these areas can be packed.

The density of magnetic storage is limited, to a large extent, by the mechanical requirements associated with reading and writing the data. Advances continue to be made that reduce the size of the read/write heads and to allow them to operate closer to the magnetic surface. The use of special thin films and metallic coatings on disks have reduced head-to-surface distances. Improvements have also been accomplished by hermetically sealing the disk and the read/write head in a single enclosure (as in the popular Winchester technology). By getting the head closer to the disk, without actually touching it, it becomes possible to pack the bits of data closer together, thereby increasing the density of storage. The storage density of hard disks has increased from about 250 bytes (or characters)-per-square-inch in 1956 to nearly 2 million characters-per-square-inch today. Even floppy disk technology, with its limiting mechanical requirements, can achieve densities of nearly a million characters-per-square-inch.

Another new technology, predicted to bring 50-to-100 fold increases in magnetic storage densities, is perpendicular magnetic recording. In a traditional magnetic medium, one can envision each of the bits of data as a tiny bar magnet aligned longitudinally in a plane with the surface of the medium. In perpendicular magnetic recording, the bar magnets can be imagined as standing on end, perpendicular to the surface of the medium. This orientation allows more bits of data to be packed into a given area. Recording densities of over 35 million bytes (or characters)-per-square-inch are predicted by 1990, allowing a 14-inch, four-platter magnetic disk to hold 5 Gigabytes (or 5 billion characters) of data.

Optical Storage

Despite the prospects for improvements in magnetic storage devices, the greatest excitement is currently being generated by optical disk technology. Whereas magnetic technology is limited by a number of physical constraints, optical technology seems to be limited only by the wavelength of light.

Optical disks typically store data as a series of spots on a light- or temperature-sensitive medium. The technology has grown out of that developed for the home-entertainment market. Both the video disks used for movies and the compact disks used for sound recordings have been adapted to store digitized information compactly and comparatively inexpensively.

"Optical-Video Disc" is the name given to computer storage adapted from home video technology. It can be used to store digital or analog information either separately or in combination. Players for these disks are now being marketed to operate in conjunction with microcomputers.

Production of optical-video disks constitutes a type of publication process wherein a master disk is made and then replicated to produce copies. The technology is ideal for preparing multiple copies of archival data. The life span of such optically recorded information is projected at more than 40 years. Optical-video disks running 12 inches in size are now being marketed with storage capacities of 2 Gigabytes (or roughly 2 billion characters).

"Compact Disc" is the name given to the audio version of this technology. When adapted for information storage, it is often referred to as "CD-ROM" or "Compact-Disc/Read-Only-Memory." These small disks are only 4.7 inches square and hold 5-600 Megabytes of data. It has been reported that IBM is developing a 2-inch version for software distribution.

Both the optical-video and compact disks are read-only storage media. Information is digitized onto a master disk and then replicated for distribution. A separate technology, called the "Optical-Digital Disc," is designed to allow data to be recorded on demand without being mastered and replicated. Recording is accomplished by a laser beam that burns pits into a special film sandwiched between two layers of Plexiglas. Once written, the disk can be read immediately. Data can be added to the disk at any time, but it cannot be erased.

The optical-digital disk offers impressive storage capacities. The "Direct-Read-After-Write" or "DRAW" disk developed by North American Philips Co. can store roughly a billion bytes of error-corrected data. A "DRAW" system currently available from Alcatel/Thomson offers 12-inch disks with a capacity of 2 Gigabytes. Projections have been made regarding the development of "juke boxes" containing 10,000 disks with a total capacity of 25,000 Gigabytes.

Combination Technologies

The three optical storage technologies we have been describing are suited for permanent storage, since data cannot be erased. This is advantageous in many applications. However, if optical technology is to compete directly with magnetic storage devices, erasable optical disks must be developed. Present efforts in this direction fall mainly into two categories. The first makes use of amorphous materials that change their state in reaction to light. Some materials are polarized by light; others change color. The second category of erasable, optical storage media combines magnetic and optical technologies. Here, heat from a focused laser beam is used to impose a magnetic orientation. This information can then be read by a polarized laser beam. Information stored in this fashion can later be erased and rerecorded.

New developments in the areas of optical and magnetic storage technology continue to improve overall storage capacity, reduce costs, and add flexibility to the way in which computers can be used. In practice, advances in storage technology have also been a driving force that stimulates development in other areas of computer technology.

III. INFORMATION PROCESSING

Major advances have also taken place in the technology that facilitates information processing and manipulation. Key areas of hardware development include 30-fold gains in processing speed; major increases in reliability; dramatic reductions in equipment size and price; and improvements in display resolution and graphics capabilities. Software developments have also been significant, including high level programming languages; powerful computer operating systems; versatile data base structures and management systems to support them; user-friendly software interfaces; and the new expert systems that begin to make use of artificial intelligence.

Of all these improvements, we think that the ones which affect microcomputer hardware and software may well have the most impact on information management to predict that microcomputer technology will soon be as ubiquitous as the telephone.

New data base management systems for micros, as well as for mainframes, are beginning to make use of relational models for data storage and retrieval. This approach is particularly well suited to the type of data likely to be collected in a national biological survey because it will allow multiple descriptors to be associated with multiple entries in the data base in a table-like arrangement. Admittedly, much of this work is still in its early stages, and most DBMSs are not yet truly relational, despite their advertising. Still, the technology is already having an impact, and there is every reason to expect that progress will continue.

IV. INFORMATION USE

The ability to use technologies in combination has also added flexibility to the ways in which data can be retrieved from a computer system and put to use. Technological developments of interest here include: telecommunications; networking; distributed and gateway systems; search software and front-end systems for retrieval; photocomposition software and hardware; and laser printing (to name just a few).

The field of telecommunications is one of the fastest growing areas in the information industry. This growth has been spurred by developments in several areas. For example, advances in fiber optics technology promise to bring improved efficiency and lowered costs to information transfer. By use of laser light and a bundle of glass strands no thicker than a finger, it is possible to transmit simultaneously over 240,000 telephone conversations. This technology shows promise for accurate transmission of computer data over long distances.

Satellite technology is also being used to reduce the cost of long distance transmissions. It has even been speculated that the data banks themselves might be put into satellites in geostationary orbits. This is not likely to happen tomorrow; but, who knows, perhaps someday your portable computer will come equipped with an antenna instead of a modem.

Getting back to the present, satellites, fiber optics, and other advances in telecommunications technology make it increasingly practical to build distributed data bases that can be accessed and maintained remotely by use of networks and gateway systems. With the aid of special "front-end" software to facilitate access, the fact that the data resides on different computers becomes almost transparent to the user. Data can be searched and/or modified remotely, or it can be copied and saved for future use locally.

Telecommunications developments facilitate the use of data in printed as well as computer-readable form. Microcomputers coupled with optical scanning devices can be used to transmit printed images to remote locations. Other developments that facilitate the use of printed output include laser printing technology, photocomposition formatting software, and graphics terminals and printers.

CONCLUSION

We have been asked to include in our talk some mention of future trends in computer technology. Of course it is easy to cite the obvious, such as continued increases in speed and capacity paired with reductions in size and price. But our crystal ball has never been particularly reliable, so we decided to look back over the past for lessons that might apply to projecting the future.

In looking back over the last 10-20 years, we were reminded that, despite the truly amazing developments in technology, it was the human element that played the major role in determining the rate at which new technology was implemented. For many years computers tended to be large and hostile devices, locked away under the charge of an elite group of professionals. It was not until they came down to a more human scale that we began to see a proliferation of computer usage and applications.

Today, computer hardware and software are becoming truly user-friendly. Over the next 5-10 years, we expect to see continued development in those areas that make computers more accessible to the end user. We also expect to see increased use of distributed systems. Dramatic improvements in processing power are being made on microcomputers as well as on mainframes. Of course it is true that advances in storage technology make it possible to build ever larger data bases in a central location. However, these same advances also make it possible to distribute copies of the data base to end users. And the same telecommunications technology that makes it possible to access a large centralized data base also facilitates access to smaller, distributed data bases.

Based on our understanding of current technology and our expectations for the future, we would like to conclude with recommendations for managing the information that will be generated by a national biological survey.

It should come as no surprise at this point that our recommendations involve a distributed information system. We acknowledge that a single centralized data store would be a much more powerful information resource than a distributed information system; however, the centralized approach also brings with it tremendous burdens for data capture, conversion, storage, and processing, to say nothing of the massive administrative and maintenance responsibilities associated with such an undertaking.

Therefore, we strongly recommend that consideration be given to adopting the concept of a "referral data base" as the mechanism for coordinating the data management aspects of the survey. A referral data base is just what its name implies. Rather than collecting all the data in one place, the referral data base 'refers' you to data available elsewhere. Under this concept a data base would be built that describes and indexes a variety of data collections available in other locations. We are sure each of you is familiar with several existing data collections that might be included. We would expect literature data bases, such as those produced by BIOSIS, to be documented in the referral data base as a source of published information on organisms.

While the data collections themselves need not be standardized, the descriptions and indexing included in the referral data base would be standardized. A model for this concept can be found in the National Environmental Referral Service (NEDRES) operated by NOAA (National Oceanic and Atmospheric Administration). Access to the distributed data collections might ultimately be provided via computer network.

In addition to recommending the development of a referral data base, we would also recommend the establishment of a central agency for the registration of organism names. This would facilitate access to the data associated with each organism and would enhance the reliability of data retrieval. This approach is already in wide use in the field of chemistry, and a mechanism for registering organism names has been under development at BIOSIS for several years.

In conclusion, we would like to say that technology today offers us a great many options for managing the information associated with a national biological survey. We believe that if we remain conservative in our approach and energetic in our execution, the technology is certainly available to support such an undertaking.

LITERATURE CITED

Allkin, R. 1984. Handling taxonomic descriptions by computer. *In*: The Systematics Association Special Vol. 26: *Databases in systematics*. Academic Press, London and Orlando.

Anonymous. (undated) Finding the environmental data you need: NEDRES (National Environmental Referral Service). National Oceanic and Atmospheric Administration. Washington, DC.

Anonymous. 1984. Special report: Laser disc storage systems. *IDP Report* 3–6, November 23, 1984.

Anonymous. 1985. Briefcase size personal computer to dominate the portable computer market. *Information Hotline* p. 10, May 1985.

Anonymous. 1985. Kodak image management system. *Information Hotline* 8–9, May 1985.

Anonymous. 1985. New disk developments: power promises for PCs. *PC Magazine* 33–34, February 19, 1985.

Anonymous. 1985. The future of personal computer communication. *Business Computer Digest and Software Review* 3(5): 1–2.

Anonymous. 1985. $1000 image scanner stores photos, text on PC disk. *Electronic Engineering Times* p. 17, April 11, 1985.

Barron, D. W. 1984. Current database design—the user's view. *In*: The Systematics Association Special Vol. 26: *Databases in systematics*. Academic Press, London and Orlando.

Beaver, J. E. 1984. New options for data crunching. *Computer Decisions* 159–164, 168.

Bisby, F. A. 1984. Information services in taxonomy. *In*: The Systematics Association Special Vol. 26: *Databases in systematics*. Academic Press, London and Orlando.

Dadd, M. N. & M. C. Kelly. 1984. A concept for a machine-readable taxonomic reference file. *In*: The Systematics Association Special Vol. 26: *Databases in systematics*. Academic Press, London and Orlando.

Freeston, M. W. 1984. The implementation of databases on small computers. *In*: The Systematics Association Special Vol. 26: *Databases in systematics*. Academic Press, London and Orlando.

Goldstein, C. M. 1984. Computer-based information storage technologies. *Annu. Rev. of Inf. Sci. Technol.* 19: 65–96.

Helm, L., J. W. Wilson & O. Port. 1985. Are U.S. chipmakers' worst nightmares coming true? *Business Week* 84a.

Kelly, M. C. & J. M. Walat. 1984. *1984 and beyond: The impact of new technology on information processing and distribution.* Presented at Council of Biology Editors Annual Meeting, Rosslyn, Virginia.

Kolata, G. 1985. Changing bits into magnetic blips. *Science* 227: 932–933.

Paller, A. 1985. The ten top graphics trends for '85. *Computer World Focus* 19(15A).

Secretariat de la faune et de la flore. 1983. *Objectifs et fonctionnement; methodologie et deontologie; programmes et publications.* Museum National d'Histoire Naturelle, Paris, ISBN 2-86515-012-6.

Semilof, M. 1985. 2,400 bit/sec modems. *On Communications* 2(5).

Shaefer, M. T. 1984. Leading-edge, high technology information devices examined by national agricultural library. Information Retrieval and Library Automation. 20(7): 1–4.

Shetler, S. G. 1974. Demythologizing biological data banking. *Taxon* 23: 71–100.

Williams, B. J. S. 1985. Document delivery and reproduction survey. *FID News Bulletin* 35(1): 8–11.

SECTION IV.
LEGISLATIVE AND HISTORICAL PERSPECTIVES FOR A NATIONAL BIOLOGICAL SURVEY

Prefatory Comments

Stephen R. Edwards
Association of Systematics Collections

There is no doubt that our federal government (both legislative and executive branches) and most, if not all, of the states have made considerable investment of funds and personnel toward gaining a better understanding of the flora and fauna of this nation.

At the federal level, a number of statutes have been promulgated that directly reference, or allude to, the broad need to conserve, understand, and document our nation's flora and fauna. These laws include: Migratory Bird Treaty Act, Lacey (=Injurious Wildlife) Act, Endangered Species Act, Marine Mammal Protection Act, Bald Eagle Protection Act, Fur Seal Act, Plant Quarantine Regulations, Noxious Weed Act, and the Animal Welfare Act. Under these federal laws over 3,000 genera and/or species of plants and animals are listed as needing special treatment.

In addition, the Endangered Species Act and the National Environmental Policy Act require all federal agencies to evaluate the "impact" of their programs on the environment and expressly prevents them from destroying specific listed habitats and species listed as "endangered."

At the state level, while only five states (Illinois, Kansas, New York, Ohio, and Oklahoma) have established biological survey programs, most states have enacted legislation designed to protect their fauna and flora. In the aggregate, over 2,500 different genera and species of animals alone are listed by the states.

It is important to note that not all federal and state legislation is designed to protect listed species. In some cases, undesirable or "pest" species are listed to prevent their introduction into the U.S. or a particular state. Nevertheless, whether a listing is designed to protect or exclude a species, there is no doubt that effective management of the species is dependent upon our understanding of the organism's distribution, reproductive biology, environmental requirements, etc. In the absence of such basic information, we are required to make decisions on the basis of educated guesses.

In addition to these publicly supported programs, an estimated 70% of the 3,800 biological collections maintained in this country are housed in tax-supported institutions. Further, nearly 200 million biological specimens and accompanying data are maintained in these 3,800 collections. Each of these collections was established to preserve, document, and communicate information about species. If 25–30% of these specimens represent native (U.S.) species, a significant resource (of between 50 and 60 million specimens) exists upon which we could further document the flora and fauna of this country.

This very brief overview of federal and state investments in acquiring species-related information documents a need and commitment for an integrated biological survey of this nation. Data are being compiled on species. Specimens are being collected and housed in museum collections. Therefore, why do we need to formalize a national biological survey?

Simply, the information that has been acquired is not accessible, and the activities I have outlined are not coordinated; therefore, the products of these activities are not available. Most often, government programs and biological collections are established for special purposes, designed and implemented to serve the interests of only a relatively few individuals—and one would have to question whether the public is considered *a priori* as a beneficiary.

To emphasize my point, I will examine the government programs and collection-related activities of which I am aware from the perspective of two parameters: Information Management and Financial Resources.

INFORMATION MANAGEMENT

Dating back to the late 1960's, with the establishment of such federal programs as the International Biological Program (IBP) and Man and the Biosphere (MABS), and through the response of federal agencies to the National Environmental Policy Act, it became clear that a crucial element necessary for management of biological information was a listing of species. Such listings would provide a means by which a vast array of species-related information (*e.g.*, locality from which the specimens were obtained, ages, reproductive conditions) could be associated, stored, and retrieved. In this manner, the data associated with a variety of specimens (from different localities) of a given species could be pooled, analyzed, and interpreted. This need led to the development of computerized systems expressly designed to store and manage biological data.

To date, I am aware of no fewer than 12 federal agencies, services, offices, and/or projects that have either developed their own computer-based, species-oriented, information management system or have developed extensive computerized listings of groups of taxa for use in conjunction with software available on their systems. These are: Office of Endangered Species, (Washington D.C.), National Park Service, Bureau of Land Management, Soil Conservation Service, Eastern Energy Land Use Team (United States Department of Interior); Forest Service (United States Department of Agriculture); Environmental Protection Agency; National Marine Fisheries Service (Department of Commerce); Department of Navy, Corps of Engineers (Department of Defense); National Laboratories (Department of Energy); and Nuclear Regulatory Commission.

The magnitude of the financial investment in these computer systems from these federal programs must be staggering when one considers the costs of personnel, computer hardware, software, space, and support facilities. And yet, the majority of these systems have either been abandoned or fall far short of performing the tasks for which they were developed.

Why Have the Majority of These Systems Failed?

I believe that one of the most important factors that has contributed to these failures is the fact that the professional systematics community was (and remains) isolated from such projects.

Given that these information management systems are developed to provide data on species according to their names, if different users employ different scientific names for the same species, then they will receive different answers.

Taxonomy is dynamic, probably changing at a rate of 5% per annum. Nomenclatural changes are communicated primarily through publications with very limited distribution designed to serve professionals working on a particular group of organisms. Finally, most qualified systematists have a relatively narrow spectrum of taxonomic expertise; therefore, it is impractical to think that each agency in the government could employ a sufficient number of qualified personnel to monitor taxonomic changes for all flora and fauna. In short, it is imposible for any comprehensive taxonomic listing to be kept up-to-date without the active participation of the professional community.

From the perspective of collections, only 21% of the specimen-related information in the U.S. is managed by use of computers. Of these collections, on the average, only data on 40% of the specimens are accessible by computer. Furthermore, only 64% of all specimens in collections have been cataloged in any manner. Less than one-half person-years is devoted to curating each collection (based on the total number of individuals associated with collections and adjusted according to the percentage of time devoted to curating the collection). Over 40% of the personnel associated with collections are volunteers or students. The average operating budget for a collection in the U.S. in 1981 was slightly over $5,600.

Given these conditions, it is not surprising that data on existing specimens are generally not available—particularly in any organized form designed to meet a public need.

FINANCIAL RESOURCES

Contrary to popular belief, considerable federal and state funds have been available to underwrite biological survey activities. According to Escherich and McManus (1983. *Sources of Federal Funding for Biological Research.* Association of Systematics Collections, 83 pages), a total of 34 federal agencies (independent of the National Science Foundation) provided nearly $3.5 billion for biological research in 1982. Because agencies were unable to provide data on support available for systematics research, the authors compiled their data on programs that substantially supported "organismic biology." The National Science Foundation (Biological Research Resources and Systematic Biology Programs) awarded approximately $3.5 million for floral and faunal surveys between 1980 and 1982.

From a different perspective, between 1977 and 1982, systematic biologists reported receiving a total of $165 million in grant/contract awards from all sources including federal programs, state and local government agencies, and private sources (Edwards *et al.*, *The Systematics Community.* 1985. Association of Systematics Collections. 274 pages). An average grant recipient received about $24,000 per year over this five-year period. Further, systematic biologists report that about 50% of their research activities are supported with personal funds.

With these facts in mind, it is clear to me that the federal and state governments, as well as the private sector, have already made a significant commitment (conceptually, financially, and in their actions) toward a national biological survey. What stands between our present condition and formal implementation of a national biological survey is a dedication of financial resources (including new funding if it is determined necessary) and coordination of existing federal, state, and private informational resources and activities toward the common goal of surveying our Nation's flora and fauna.

Federal Legislation and Historical Perspectives on a National Biological Survey

Michael J. Bean
Environmental Defense Fund

Abstract: This paper examines some of the federal laws and programs that either require biological survey-type undertakings or that would be significantly aided by such efforts. Also recounted are some of the historical efforts of federal agencies to undertake biological surveys or establish networks of protected areas for ecological research on federal lands.
Keywords: Federal Legislation, Historical Perspectives, Biological Survey.

INTRODUCTION

Though the term "biological survey" apparently implies somewhat different things to the many different people who use it, the one common characteristic in most definitions is the comprehensive identification and description of all the biota of a region. There is currently no federal legislation requiring that a "national biological survey" of the U.S. be undertaken. Nevertheless, some of the purposes that such a survey might serve can probably be substantially achieved through a variety of existing laws enacted primarily for other purposes.

Historically, the conservation of living resources has been the responsibility primarily of the states rather than the federal government. The federal conservation role still remains rather limited, focusing mainly on migratory birds, marine mammals, and endangered species; the states, though in theory responsible for everything else, have concentrated even more narrowly on game species. More recently, the assertion of a federal interest in preserving general environmental quality—as opposed to specific living resources—has given the federal government a new set of responsibilities for which biological survey data could be of major value. This paper reviews the history of the first biological survey efforts of the federal government, then examines some of the current legislation upon which a survey could be built today.

HISTORY

For nearly half a century, there was within the federal government an agency known at various times as the Division or Bureau of Biological Survey. The agency's origins can be traced to an 1885 congressional appropriation of $5,000 to the Department of Agriculture's Division of Entomology "for the promotion of economic ornithology, or the study of the interrelations of birds and agriculture." (Act of March 3, 1885, 23 Stat. 353, 354). The next year, Congress sought to expand and perpetuate that effort by creating a Division of Economic Ornithology and Mammalogy in the Department of Agriculture. Dr. C. Hart Merriam of the American Ornithologists' Union became the first Chief of the new Division, which was charged with investigating "the food-habits, distribution, and migrations of North American birds and mammals in relation to agriculture, horticulture, and forestry." (Act of June 30, 1886, 24 Stat. 100, 101). A decade later, both the name and statutory duties of the agency were broadened; it became the "Division of Biological Survey" and was given a general charge to carry out "biological investigations, including the geographic distribution and migration of animals, birds, and plants." (Act of April 25, 1896, 29 Stat. 99, 100).

The 1896 expansion of the agency's statutory duties only codified what the agency had in fact already been doing, for by that time it had commenced major collecting efforts not just for birds and mammals, but also for reptiles and amphibians in the U.S. and elsewhere in the Western Hemisphere. Major scientific surveys had been completed or were in progress in many of the western states, Alaska, Canada, and Mexico by the turn of the century.

The early years of the new century were transitional ones for the agency. In 1905, its name was changed again to the Bureau of Biological Survey. More importantly, it acquired new duties to aid in the enforcement of state wildlife laws (1900), manage the first components of what would become the National Wildlife Refuge System (1903), conserve migratory birds (1907), and control predatory and noxious animals (1909). These new duties quickly transformed the major mission of the agency from conducting survey-oriented research to active management of living resources and their habitats. In 1939, an executive reorganization transferred the agency to the Department of the Interior, where in the following year it was merged with the Bureau of Commercial Fisheries (transferred from the Commerce Department) to become the U.S. Fish and Wildlife Service. By World War II, statewide faunal survey work had ceased, though the Fish and Wildlife Service was at least partially responsible for the production of three recent state surveys: "Mammals of Maryland" (Paradiso, 1969), "The Bird Life of Texas" (Oberholser, 1974), and "Mammals of New Mexico" (Findley et al., 1975).

The extensive collections that were begun during the early surveys are now under the curatorial care and management of the "Biological Survey Section" of the Fish and Wildlife Service's Denver Wildlife Research Center. They are in the National Museum of Natural History in Washington, D.C. With a staff of only 16 and an annual budget of some $700,000, today's "Biological Survey Section" has been called "a struggling, dispirited bureaucratic waif far from the center of power." (Rensberger, 1985).

CURRENT LEGISLATION

To date, there has been no public policy, clearly embodied in any federal statute, to carry out a general biological survey of the nation. However, more limited and specific surveys or survey-like undertakings have often been essential to the proper carrying out of other policies. Thus, a variety of federal agencies, operating under different statutory mandates, can carry out significant pieces of a biological survey. Together, their efforts, if properly directed, could contribute a substantial share of a national survey effort.

1. NEPA and Related Laws

The single most pervasive and influential federal environmental law of the modern era is the National Environmental Policy Act of 1969 (42 U.S.C. 4321 *et seq.*). It applies to all federal agencies and affects the way in which each of them does its business. NEPA articulates several major policy goals, among them to "preserve important historic, cultural, and natural aspects of our national heritage, and maintain, wherever possible, an environment which supports diversity and variety of individual choice." (42 U.S.C. 4331(b)(4)) The principal means by which this goal is to be served is to require each federal agency to prepare a "detailed statement" (commonly known as an environmental impact statement) for each major action significantly affecting the quality of the human environment. These statements are to describe "the environmental impact of the proposed action" and its alternatives. (42 U.S.C. 4332(2)(C))

The effective realization of the goals set forth in NEPA would clearly require a substantial body of baseline environmental information like that which a biological survey could provide. In practice, however, the process of preparing an environmental impact statement rarely begins with anything like a comprehensive biological survey. The reason is that NEPA's command to describe the environmental impact of a proposed action has been qualified, by administrative and judicial interpretation, to limit that duty to "important" impacts. "Importance", in turn, can most readily be discerned by reference to those federal and state laws expressing some clear public policy relating to living resources. Thus, to the extent that NEPA impact statements address impacts on specific living resources, they almost always concentrate on migratory birds, endangered species, marine mammals, and commercially or recreationally valued fish and wildlife, precisely because these are the types of living resources for which clear public policies have been developed and expressed in state and federal conservation legislation. In short, a comprehensive survey of the tiger beetle fauna in the area of a proposed major federal development is simply not likely to be viewed as necessary, absent some endangered species or other clear public policy consideration. Thus, NEPA as a tool with which to build a national biological survey is not likely to be very effective.

The same general observation can be made with respect to a number of other federal laws that, like NEPA, try to influence development decisions by ensuring that those who make such decisions are fully informed as to the environmental consequences of their decisions. The Fish and Wildlife Coordination Act (16 U.S.C. 661 *et seq.*) is the oldest and one of the most important of these. It was originally passed in 1934 in response to the destruction of valuable anadromous fishery resources as a result of the damming of rivers, and had among its limited purposes encouraging the construction of fish ladders to aid those resources. As subsequently amended, it seeks to ensure that "wildlife conservation...receive equal consideration and be coordinated with other features of water-resource development programs," including the permit programs for the filling of wetlands and the discharge of pollutants into water under the Clean Water Act. (16 U.S.C. 661). The mechanism through which the Act works is consultation between the federal resource development or permitting agency and the U.S. Fish and Wildlife Service and its state-level counterparts.

Once again, in practice, the focus of attention in the actual implementation of the Fish and Wildlife Coordination Act is not on tiger beetles, caddisflies, or flatworms. Rather, it is on game fish, migratory waterfowl, furbearers, and endangered species. Indeed, the typical currency by which the impacts of proposed actions are measured under the Coordination Act are "hunter-days" or "angler-days."

To some extent this narrow focus of concern under NEPA, the Coordination Act, and similar laws may be broadened as a result of the growth of nongame conservation programs at the state and federal levels. In 1980 Congress passed a law to encourage states to develop conservation programs for wildlife not hunted or caught for commercial purposes. The Fish and Wildlife Conservation Act of 1980, as it was called, was to have provided federal funding in support of such state programs. (16 U.S.C. 2901 *et seq.*) To date, no federal funds have been appropriated, although a number of states have nevertheless established active nongame conservation programs funded entirely with state revenues. The federal law requires that state programs, to qualify for federal financial assistance, "provide for an inventory of the nongame fish and wildlife...that are...valued for ecological, educational, esthetic, cultural, recreational, economic, or scientific benefits by the public." (16 U.S.C. 2903)

One survey-like activity that has grown out of the Fish and Wildlife Coordination Act and the migratory bird conservation programs of the U.S. Fish and Wildlife Service is the undertaking by the Service of a revised "National Wetlands Inventory." The first such inventory was completed in 1954, and the revision began two decades later. Still some years from completion, the purpose of this effort is to classify wetland types and identify their distribution and extent. The National Wetlands Inventory does not attempt to be a comprehensive "biological survey" of each wetland area or type. However, the Service's Office of Biological Services has prepared "community profiles" for wetland areas, including New England high salt marshes, southern California coastal salt marshes, and south Florida mangrove swamps, atlases of coastal waterbird colonies for the Atlantic, Pacific, and Alaskan coasts, and a myriad of other similar ecological studies have been undertaken by that Office's "Coastal Ecosystems Project." Similar ecological studies pertaining to areas likely to be most affected by coal development have been undertaken by that Office's Eastern and Western Energy and Land Use Teams.

2. Federal Land Management Legislation

Most federal land managing agencies, including such major agencies as the U.S. Forest Service, Bureau of Land Management, National Park Service, and Fish and Wildlife Service, are subject to resource inventory requirements for the lands under their jurisdiction. Together, these lands comprise nearly a third of the total land area of the U.S., although their geographic distribution is heavily skewed to the West.

The U.S. Fish and Wildlife Service, successor agency to the old Bureau of Biological Survey, is responsible for administering the National Wildlife Refuge System. The Refuge System is one of four major federal land-management systems. It is the only one that has as its paramount purpose the conservation of living resources. Congress has not clearly articulated a detailed statement of purposes for the Refuge System, but others, both within and outside the Service, have often expressed the view that it should eventually encompass representative samples of all the major ecosystem types or life zones in the U.S., so as to provide protected habitat for most, if not all, of the native vertebrate species and for many invertebrates. Nothing in the federal laws pertaining to the Refuge System requires that the Service carry out any type of biological survey, yet doing so would clearly aid the purposes described above.

In addition to the National Wildlife Refuge System, three other major federal land systems exist for which the governing laws may be relevant to any national biological survey effort. These are the National Park System, administered by the National Park Service of the Department of the Interior; the National Forest System, administered by the U.S. Forest Service of the Department of Agriculture; and the National Resource Lands, administered by the Bureau of Land Management of the Department of the Interior. The first of these is managed to conserve nature and provide public recreation. The latter two are managed for "multiple use" purposes that, though they include conservation, have traditionally emphasized commodity production.

Both of the multiple-use land systems are managed under federal statutes that require the preparation and periodic revision of "resource inventories". The National Forest Management Act requires the Secretary of Agriculture to "develop and maintain on a continuing basis a comprehensive and appropriately detailed inventory of all National Forest System lands and renewable resources". The limitation of this mandate to "renewable resources" suggests that only those living things with some commercial or recreational use might be included within the scope of the inventory. Though the very clear emphasis of the inventory work done to date has been precisely on such species, it has not excluded others, such as invertebrates that are protected as threatened or endangered by federal or state law, and those "known or likely to be particularly sensitive to the management of" Forest Service lands (U.S. Forest Service, 1980, p. 165).

The Federal Land Management and Policy Act, the basic federal legislation governing the management of lands under the jurisdiction of the Bureau of Land Management, has a similar inventory requirement. That statute requires BLM to maintain an inventory of its lands and "their resource and other values...giving priority to areas of critical environmental concern". The idea of an "area of critical environmental concern" or "ACEC" was a new statutory concept intended to embrace areas "where special management attention is required to protect ...important...fish and wildlife resources or other natural systems or processes".

ACECs are linked to a much earlier effort by federal land-managing agencies to identify and protect areas under their jurisdiction with particular ecological or research value. In 1927, the Forest Service established the first of what was to become an extensive, informally connected system of "research natural areas". In the following decade, the Park Service established 28 similar "research reserves" in ten national parks. Later, the Park Service took the lead in advocating a coordinated effort among all federal land-managing agencies to inventory and protect those areas under their jurisdiction that had either representative or unique characteristics of value for research or educational purposes. That effort was aided by a 1965 "Special Message to the Congress on Conservation and Restoration of Natural Beauty", in which President Johnson directed that a study be undertaken "to recommend the best way in which the federal government may direct efforts to advancing our scientific understanding of natural plant and animal communities and their interaction with man and his activities." The following year the Federal Committee on Research Natural Areas was formed (The Nature Conservancy, 1977).

The Federal Committee on Research Natural Areas was an informal, administratively created committee comprised principally of representatives of federal land-managing agencies. It had no statutory mandate, no appropriated funds, and no staff. Its purpose was to encourage the development of a system of specially designated and protected research and education areas. At first, the intention was to allow natural ecological processes to govern in such areas. Later, the system added areas where experimental, manipulative management was to be carried out. The name of the committee evolved as well. In 1974, the committee was renamed the Federal Committee on Ecological Reserves. It formally adopted and published a charter by which it announced its objectives, which included expansion of the existing system to include additional federal lands as well as state, local, and private lands, and developing guidelines and criteria for management of the areas of the system. In 1977, the Committee published a comprehensive directory of federal research natural areas that included nearly 400 areas totalling more than four million acres (Federal Committee on Ecological Reserves, 1977). Though still other areas were added later, by 1980 the Committee was moribund. It last met in December 1979, but has not been formally disbanded. The special areas it helped establish continue.

3. The Federal Endangered Species Program

In most federal conservation and environmental laws, biological survey-type information would be valuable primarily for its utility in furthering the conservation goals that pertain to a relatively few species, among them migratory birds, marine mammals, and commercially and recreationally valued wildlife. The most significant exception to this rule is the federal Endangered Species Act. Administered principally by the Fish and Wildlife Service (the National Marine Fisheries Service of the Commerce Department is responsible for marine organisms), the Endangered Species Act is probably most directly relevant to renewed concern for a national biological survey because of its geographic and taxonomic scope. Its purpose is to provide a program for the conservation of all species—vertebrate, invertebrate, and plant—that are now threatened with extinction or are likely to become so within the foreseeable future. The Act is noteworthy, among other reasons, because it reaches beyond vertebrates and commercially valuable shellfish; virtually no earlier federal conservation law did so (Bean, 1983).

To identify and protect those species eligible for protection under the Act, some reasonable systematic method of inventorying and assessing the status of plants and animals is essential. Because very little was known about the conservation status of plants when the Act was passed in 1973, the Act directed the Smithsonian Institution to carry out a special study to identify plants that might need protection under the Act. The Smithsonian study, completed in 1975, identified roughly 3,000 species of vascular plants from the U.S. that appeared to be in jeopardy of extinction (Smithsonian Institution, 1975).

The comprehensive effort that the Smithsonian study represented provided a model for a later review by the Fish and Wildlife Service itself of about 400 species of vertebrates (in 1982) and 1,000 species of invertebrates (in 1984) that it believed, on the basis of literature reviews and information from professionals in the field, might warrant the Act's protection. From these three separate reviews, the Fish and Wildlife Service has refined a current list of about 3,000 species that it regards as active "candidates" for future listing. In practice, whenever a new species, particularly a vertebrate, is described, it has a good chance of becoming a candidate for future listing. The fact that it has been undescribed heretofore is often indicative of its limited distribution or numbers.

The Endangered Species Act confers protection only on those species that have actually been listed as threatened or endangered, of which there are currently about 350 from the U.S. "Candidate species" receive no legal protection; indeed, the very concept of candidates was an administrative invention of the Fish and Wildlife Service rather than a statutory creation of Congress. Nevertheless, the Act provides some important opportunities to gather additional information about the distribution, abundance, and general conservation status of these species. The Fish and Wildlife Service itself, because it has the candidate species under active consideration for future listing, endeavors continuously to gather new data about them so as to expedite their listing and protection if their conservaton status declines or to remove them from the candidate lists if they prove to be less imperiled than originally thought. Because of perceived deficiencies in that effort, Congress is currently considering, and is likely to enact, amendments that would establish a statutorily mandated monitoring program for candidate species.

Another means the Act provides to gather information about the distribution, abundance, and status of candidate species is through cooperation with, and financial incentives to, the states. One of the principal purposes of the Endangered Species Act was to stimulate active state programs for the conservation of species other than the game animals and furbearers that had long been the almost exclusive preoccupation of most state wildlife conservation agencies. The mechanism enabling this is Section 6 of the Act, which offers partial federal financial assistance to those states that establish conservation programs meeting certain minimum federal standards. Among the requirements for states seeking to qualify for such assistance is that they be "authorized to conduct investigations to determine the status and requirements for survival of resident species" of plants or animals. Currently, more than forty states have established qualified programs for wildlife and nearly twenty for plants. Although the amount of federal funds available to aid in the implementation of state programs is small (currently about $5 million annually), one of the frequent purposes for which such funds are supplied to conduct status surveys of some or all of the candidate species in the state. Congressional interest is currently quite high in bolstering the cooperative programs encouraged by Section 6. Despite the general restraints on spending currently prevailing in Congress, the House of Representatives this year voted to double the spending ceiling that currently applies to federal support for state endangered species programs, up to $12 million annually.

Federal agencies other than the Fish and Wildlife Service also may need to carry out extensive survey work as a result of the Endangered Species Act. Section 7 of the Act imposes a significant and enforceable (through citizen lawsuits) duty on all federal agencies to ensure that actions they authorize or carry out do not jeopardize the continued existence of any threatened or endangered species. For most federal projects, the inital step in compliance with that duty is for the federal agency proposing the project to conduct a "biological assessment" to identify any listed species that is likely to be affected by the project. Such biological assessments typically involve on-the-ground surveys in the area of the proposed project for not only listed species but also for candidate species. The results are reported to the Fish and Wildlife Service, which uses them to evaluate the proposed project and to refine its evaluation of the conservation status of the species.

In these various ways, the Endangered Species Act provides opportunities to gather a substantial amount of survey-like information about the significant number of rare plant and animal species. Though conducting a biological survey is not explicitly mandated by the Act, effectively achieving the Act's purposes cannot be accomplished without a comparable undertaking.

CONCLUSION

The various species conservation laws, land management statutes, and other environmental laws of the federal government provide a framework upon which a concerted biological survey effort could be built. The one law that clearly has the conservation of biological diversity as its central aim, the Endangered Species Act, has thus far lacked the resources to carry out successfully its fundamental purpose. However, the work done under that Act, and in particular the identification of a large number of plants and animals in the U.S. that are legally unprotected but nevertheless in peril of extinction, provides an opportunity for other agencies administering other programs to coordinate those programs with the Endangered Species Program in ways that would both substantially aid the conservation of biological diversity and achieve many of the purposes of a national biological survey.

ACKNOWLEDGEMENT

The author thanks Dr. A. L. Gardner of the U.S. Fish and Wildlife Service for his suggestions concerning the historical section of this chapter.

LITERATURE CITED

Bean, M. 1983. *The evolution of national wildlife law.* Second Edition. Praeger, New York. 449 p.

Federal Committee on Ecological Reserves. 1977. *A directory of research natural areas on federal lands of the United States of America.* U.S. Forest Service, Washington, DC. 280 p.

Findley, J. S., A. H. Harris, D. E. Wilson, & C. Jones. 1975. *Mammals of New Mexico.* University of New Mexico Press, Albuquerque, xxii + 360 p.

Oberholser, H. C. 1974. *The bird life of Texas.* University of Texas Press, Austin, 2 volumes, xxviii + 1069 p.

Paradiso, J. L. 1969. *Mammals of Maryland.* North American Fauna, No. 66, iv + 193 p.

Rensberger, B. 1985. Biological survey an orphan at 100. *The Washington Post*, July 1, 1985, p. A3.

Smithsonian Institution. 1975. *Report on endangered and threatened plant species of the United States*, H. R. Doc. No. 94-51, 94th Cong., 1st Sess.

The Nature Conservancy. 1977. *Preserving our natural heritage.* Government Printing Office, Washington, DC. 323 p.

United States Forest Service. 1980. *An Assessment of the Forest and Range Land Situation in the United States.* Washington, DC. 631 p.

State and Private Legislative and Historical Perspectives, with Comments on the Formation of a National Biological Survey

Paul G. Risser
Illinois Natural History Survey

Abstract: There are five active state biological surveys today, though other states have had or are planning biological surveys. In those states currently without biological surveys, frequently some or all of the survey functions are distributed among several state agencies. A number of private organizations also contribute to the functions of biological surveys, but rarely at a comprehensive national scale. This paper summarizes the functions of existing state and private organizations that contribute to biological survey activities, and then specifies 12 expectations of a national biological survey. Finally, a four-step process is recommended for defining and implementing a national biological survey.
Keywords: Species Inventory, Habitat Inventory, Collections, Synthetic Publications, Information Clearinghouse, Plan.

INTRODUCTION

Considerable effort has been expended toward biological survey activities at the state level, although the resulting continuing programs are quite diverse among the various states. This effort has been exerted by state and private organizations, frequently operating in concert, but the preponderance of activity has come from state agencies and institutions. In the following text, I will examine the types of programs developed by state and private organizations, evaluate ways in which these activities could be enhanced by implementation of a national survey, ways in which these activities could contribute to a national survey and, finally, pose specific recommendations about the formulation of a national biological survey.

STATE ORGANIZATIONS

Though there are notable exceptions, most states do not have organized biological surveys (National Wildlife Federation, 1984; The Nature Conservancy,

1977). Illinois, Kansas, New York, Ohio, and Oklahoma have active biological surveys today, and other states, e.g., Wisconsin, have had active biological surveys in the past. The existing state biological surveys have broad mandates to study and report on the flora and fauna of the state, but in all cases, major emphases have been on conducting inventories of species and habitats, building and maintaining biological collections, producing monographic and synthetic publications, providing management recommendations, predicting consequences of human and natural disturbances, and generally operating as a clearinghouse for data and as a central location for obtaining advice and information. Except where state biological surveys exist, several agencies within one state may participate in some or all of these activities.

PRIVATE ORGANIZATIONS

In addition to private colleges and universities, a number of private organizations participate in activities related to a biological survey. These can be conveniently categorized into three types of organizations. The most conspicuous include those private organizations that seek to preserve habitats, such as The Nature Conservancy. The Conservancy not only protects habitats, it also maintains a large data base on species occurrences in various states and summarizes this information at the national level. The Heritage Programs of the Conservancy now operate in 35 states and comprehensively organize information on species and habitats. A second category of private organizations is that of research institutions, frequently associated with special reserves. Examples include the Tall Timbers Research Station in Florida and the Max McGraw Wildlife Foundation in Illinois. These institutions support research on their own property or via monetary support for specific projects. The third type consists of a diverse group of organizations whose general objective is to enhance environmental quality, in this case by supporting inventories, habitat acquisition, or stimulation of policies that ensure high quality habitat and maximum biological diversity. Examples of the latter include the Audubon Society and ad hoc groups formed to address local issues or proposed projects.

Except for The Nature Conservancy and perhaps some conservation groups such as Ducks Unlimited, most biological survey programs conducted or sponsored by private organizations are not designed to be consistent among states. Moreover, these programs frequently focus on a single issue and do not attempt to examine biological resources comprehensively over long time frames within political or regional boundaries.

EVALUATION OF STATE AND PRIVATE BIOLOGICAL SURVEYS

To evaluate the objectives and programs of state and private biological survey efforts, I have chosen to use the six fundamental activities of existing state biological surveys (Risser, 1984).

1. Conducting Inventories of Species and Habitats

Most of this activity is performed by state or private colleges and universities, though under these circumstances there is usually no long-term master plan. Some museums maintain field inventory programs, though these are customarily focused on certain taxa and in certain areas. Many game and fish departments also inventory species and habitats, and with the advent of non-game species programs, these inventories have begun to involve a wide variety of species, although the emphasis is still on game species. Many states have a natural resources agency that maintains biological inventories, although the inventories are usually general and may not include the actual data.

2. Building and Maintaining Collections

Though there are a few private collections, most collections of biological specimens are housed in colleges, universities, and city or county agencies. Support for these collections comes from state sources, which are frequently supplemented by external research or environmental evaluation funding.

3. Producing Monographic and Synthetic Publications

Monographic and synthetic publications are usually produced by faculty associated with colleges and universities or by scientists working in museums or biological surveys. State agencies frequently publish descriptions of biological resources, but these are usually general and at the layperson level.

4. Providing Management Recommendations

Recommendations about how to manage biological resources are made by state biological surveys but most often by game and fish departments and natural resource agencies. Colleges and universities may contribute to this role, but these recommendations are usually the result of a specific study.

5. Predicting Consequences of Human and Natural Disturbances

Environmental evaluations are conducted by several private and state organizations. Faculty at colleges and universities participate in this activity, both as consultants to private firms and as participants in programs sponsored by state agencies. Some private consulting firms maintain data bases, but this information is usually secondary in nature. Evaluation of natural disturbances is usually an academic process conducted in colleges and universities. Museums contribute to the evaluation process by supplying information from current and past collections.

6. Acting as a Clearinghouse for Information and a Center for Advice

The clearinghouse role is fragmented in most states. Even for large, well- funded collections, obtaining summarized information is difficult. State agencies usually have a specific range of topics under their jurisdiction, so the user must communicate with several sources to find required information about a range of biological resources. Furthermore, this information is usually not compatible in format, duration, or quality. Other than The Nature Conservancy, private organizations do not generally serve an information clearinghouse function.

EXPECTATIONS OF A STATE BIOLOGICAL SURVEY

Only five state biological surveys are active, though the same functions are frequently performed by several agencies in other states. Even the existing state biological surveys differ in scope and size. Thus, summarization of expectations of a state biological survey demands the examination of many state agencies and, to a smaller extent, private organizations. In brief and generalized form, organized state biological surveys would be expected to perform the following roles, perhaps in conjunction with other state agencies and institutions.

1. Act as a focal point to bring biological resource issues to the attention of the public and decision makers.
2. Maintain collections of biological specimens and constantly assess the status of species in the state.
3. Identify and document alarming trends or sudden changes in species population numbers and geographic or habitat distributions.
4. Encourage scientists and interested lay people to investigate and understand the state's biological diversity.
5. Establish priorities for research and management programs aimed at the biological resources of the state.
6. Provide biological information to be used in refining and employing environmental quality indicators.
7. Provide comparative information on species and ecosystems so that local and state entities can assess the value of their biological resources.
8. Provide specific products such as lists of experts, species distributions, ecosystem processes, identification manuals, taxonomic and ecologic data, data indices and summaries, and scientific and public information publications.
9. Provide information for responding to legal mandates such as habitat protection, resource inventories, and threatened and endangered species.
10. Train students and provide a repository for the results of research.
11. Provide information to the industry-business community to increase their appreciation of the value of biological resources and processes.
12. Act as a central clearinghouse of information and advice to increase the efficiency of communication and the quality of decisions about the state's biological resources.

RELATIONSHIPS BETWEEN STATE BIOLOGICAL SURVEYS AND A NATIONAL BIOLOGICAL SURVEY

The listed expectations of state biological surveys are many of the expectations that should be realized by a national biological survey. Thus, except for the coordination role necessary at the national level, 50 active state biological surveys could contribute the essence of a national biological survey. That is, in the dispersed mode, a national biological survey could consist of active state biological surveys and an information coordination function that would permit the addressing of issues at the national level. Both federal and state funding would be necessary to develop state surveys throughout the country, and federal funding would be required for the national coordination activities.

A national biological survey is additionally necessary if the state surveys are to influence decision-making at the national level. That is, national priorities and policies usually depend upon a comprehensive assessment from several or all states. This can be accomplished only by a nationally coordinated effort. On the other hand, the value or jeopardy of a state's biological resources can be evaluated only in the context of the nation, so a national survey is necessary if the full potential of the state biological surveys is to be realized. To be certain of long-term stability, state biological surveys require a core of consistent state funding. Part of these state programs can be maintained by research funding from various sources. However, the fundamental objective of biological surveys is to maintain long-term records and collections—activities that are not amenable to project-by-project funding.

RECOMMENDATIONS

In formulating recommendations, it is important to recognize several fundamental issues. First, a national biological survey would relate to many existing organizations. Though this paper considers state and private agencies, several federal agencies also contribute to the substance of a national biological survey. Thus, formulation of a national organization immediately precipitates some concern about the possible impacts on existing organizations. Second, the prevailing national budget climate is not particularly conducive to increased funding of widely visible programs, especially as compared to social support programs. Third, the concept of centralization is regarded with suspicion by many individuals and organizations. So, there are many unresolved basic considerations, such as whether a survey would be a loose amalgamation of existing organizations, an entirely new cabinet-level department, or an added responsibility for an existing organization. Fourth, the magnitude of the task of conducting a national biological survey would seem so large that the basic feasibility would be questioned by some. Fifth, the establishment of a national survey will require successful political persuasion. This is an activity in which the biological community has not demonstrated consistent accomplishments.

Establishing a national biological survey is an important endeavor that, because of the above issues, could easily become disorderly and unsuccessful. The ultimate organization must ensure that the expertise and data are closely associated but that there is sufficient national coherence to obtain the benefits of a nationwide program. Rapid, unplanned initiatives will undoubtedly produce sufficient adverse reactions so as to jeopardize the process and, therefore, the ultimate benefits. Thus, I believe that the scientific community should invest a year or two in planning the orderly development of a national biological survey. My own notion is that a national biological survey should build on the existing organizations and institutions but develop enough central structure to coordinate information. Furthermore, the task is large and important; ergo, substantial money and effort should be devoted to such a national biological survey. However, there are a plethora of possible scenarios, and these should be thoughtfully considered by the scientific community.

Specifically, I recommend the following procedure:

1. That a planning-steering committee be established, perhaps as proposed by the American Institute of Biological Sciences.
2. That the steering committee accomplish the following tasks:
   A. Articulate a tentative set of objectives for a national survey.

B. Organize a series of workshops to address the status of current activities related to a national survey, evaluation of amounts and quality of existing data and collections, data management, potential users of the program, organizational structure; and an agenda for a national congress.
C. That a national congress be conducted on the basis of deliberations of the workshops. This congress would include a representative group of participants and users and would be expected to produce a clear mandate supportable by the scientific and user community.

3. That a plan of action be developed from the products of the workshop, to include an organizational structure and a strategy for obtaining the necessary political support.
4. That the national biological survey be implemented.

LITERATURE CITED

National Wildlife Federation. 1984. *Conservation directory, 1984.* 29th Edition. Washington, DC. 297pp.

The Nature Conservancy. 1977. *Preserving our natural heritage. Volume II. State activities.* U.S. Government Printing Office. Washington, DC. 671pp.

Risser, P. G. 1984. *Illinois Natural History Survey Annual Report. Highlights of 1983-1984.* Illinois Natural History Survey. Champaign, IL. 35pp.

SECTION V.

INTERNATIONAL PERSPECTIVES

Prefatory Comments

Lloyd Knutson
Biosystematics and
Beneficial Insects Institute

Ke Chung Kim
The Pennsylvania State University

An enterprise as important as a national biological survey has international significance and relationships, just as most "national" collections are of international consequence. In addition to a biological survey's linkages with different disciplines, institutions, and other national endeavors, this symposium also considered linkages of a U.S. biological survey with other geographical areas and with other countries. We were fortunate in having at the symposium key persons who are extensively involved in biological survey activities in other countries: Peter B. Bridgewater, Australia; Hugh V. Danks, Canada; and José M. Sarukhán, Mexico.

What are the international relationships? First of all, species do not, of course, respect political boundaries. A biological survey of the U.S. will involve many species that also occur in Canada, Mexico, and other countries. The geographical distributional facts of life mean that a U.S. national biological survey will be fundamentally related to survey activities and systematics research in other countries, particularly Canada and Mexico.

Not only is almost any "national" survey, perforce, a part of "international" survey activities, but also the limited taxonomic expertise in any one country demands a high degree of international cooperation. There are taxonomic specialists in various countries with unique expertise in the plants and animals whose distributions include the U.S., and that expertise may be lacking in the U.S. Hopefully, such expertise can be brought to bear on a U.S. national biological survey.

There are various successful models in other countries for a national biological survey. As a biological survey is developed in the U.S., we need to learn from these models. Both the *planning* experience and experiences in *conducting* a national biological survey in other countries will be useful to the U.S. For example, it is interesting to note, in the case of both the Canadian and Australian surveys described in subsequent chapters, the close ties with ecological and environmental interests and the nature of these surveys as primarily initiators and supporters of survey work, rather than being institutions directly conducting survey activities.

The kinds of taxonomic studies that will provide the range of needed information are best conducted on a regional or worldwide basis, covering the entire geographical range of the taxon. A U.S. national biological survey will best be served by such comprehensive monographs and revisions with sound classifications. Obviously, development of this kind of research is enhanced by strong international linkages.

One of the major unresolved issues facing a U.S. survey is the survey's relationship to the strong interest in and need for work in the tropics. Emphasis on a U.S. survey should not, need not, detract from work in the tropics. *Both* are highly significant scientific and societal needs. Although the rate of loss of biotic diversity is faster in the tropics than in the temperate regions, thus arguing for a certain priority, the impact of biotic loss due to demographic and environmental causes is immediate to our daily lives in temperate areas. Major areas in both regions appear doomed, and biological surveys in both temperate and tropical areas have a high parallel priority. An exchange of letters in regard to the Central American insect fauna (Janzen, 1985; Adams, 1985; Miller, 1986) points up the needs for work on the tropical fauna and the potential relationships with a biological survey in the U.S.

As planning for a U.S. survey proceeds, the broad range of *international* concerns and experience will need to be considered. The increasingly broadened international perspective of the Association of Systematics Collections (typified, for example, by the nature of the Association's 1986 annual meeting) should prove useful. The enhanced value of activities conducted on an international scale were perceptively analyzed by Stuessy (1984). We are, in fact, seeing the value of a broader, international perspective in areas such as biosystematic services in entomology, where an International Advisory Council, recently formed as a result of a symposium at the XVIIth (1984) International Congress of Entomology (Kim, in press), is beginning to bear fruit.

LITERATURE CITED

Adams, R. McC. 1984. Letter to D. H. Janzen, in *Commentary*. Degradation of tropical forests: a dialogue. *Bull. Entomol. Soc. Amer.* 31: 12–13.

Kim, K. C. In press. International Advisory Council for Biosystematics Servies in Entomology. *In*: L. Knutson, K. M. Harris, and L. M. Smith (eds.) *Biosystematic Services in entomology. Proceedings of a symposium held at the XVIIth International Congress of Entomology, Hamburg, Federal Republic of Germany, August 20-26, 1984*. Agric. Res. Serv., U.S. Dept. Agric.

Janzen, D. H. 1984. Letter to R. M. Adams in *Commentary*. Degradation of tropical forests: a dialogue. *Bull. Entomol. Soc. Amer.* 31: 10–12.

Miller, D. R. 1986. Letter in *Commentary. Bull. Entomol. Soc. Amer.*

Stuessy, R. F. 1984. The organizational development of the systematic biology community. *ASC Newsletter* 12: 49–53.

The Australian Biological Resources Study: 1973-1985

P. B. Bridgewater
Australia Bureau of Flora and Fauna

Abstract: Since 1973, the Australian Biological Resources Study (ABRS) has improved taxonomic and ecological knowledge of plant and animal distribution through provision of a grants program and through development of national databases and publications. Institutions in states and territories are the major source of taxonomic activity, which ABRS has integrated into a national framework. Data available suggest a 50% increase in taxonomic productivity since the inception of ABRS.

Apart from species catalogues, floras, and faunas, ABRS is helping develop a series of databases that will be available as a national source of information on the taxonomy and distribution of Australian biota. Part of that program includes the development of techniques to use such data in active conservation and land management.

Keywords: Australia, Floras, Faunas, Databases, Species Mapping, Vegetation Mapping, Catalogues.

INTRODUCTION

Unlike most institutions that maintain collections, the Australian Biological Resources Study (ABRS) has no statutory obligation to collect and curate—rather it is an organization charged with answering the questions:

–What sort of animals and plants do we have in Australia?

–Where are these animals and plants found?

Clearly answers to these questions are of enormous importance and interest to conservation and land use managers. Indeed, it is true to say that without adequate answers to those questions, any land management or conservation decisions are likely to be less than fully effective. The role of the ABRS then, is to provide the structure necessary for continued improvement to the sciences of taxonomy, biogeography, and descriptive ecology, as one element in the development of sound conservation strategies. This point is well emphasized in the Australian National Conservation Strategy (1984), where, under the heading of Priority National Actions, Research, the following are listed:

a. Strengthen research efforts to improve knowledge of the different life support systems, of their capability for being used for different purposes and of the management required to sustain their capability for those uses.

b. Improve national coordination of environmental research so that effort is better directed toward agreed priorities, duplication is avoided and adequate communication exists between research agencies.

c. Improve taxonomic and ecological knowledge of plant and animal species and their distribution, impacts and interrelationships.

AIMS AND OBJECTIVES OF THE ABRS

Goals of the ABRS, reflecting Government policy in this area, are to:

1. Coordinate all work aimed at collecting, describing and classifying Australian plants and animals and determining their distributions (including vegetation mapping) and to maintain liaison with international bodies engaged in similar activities.
2. Establish priorities for taxonomic and biogeographic research to facilitate regular publication of a systematic series of flora and fauna handbooks.
3. Develop a national biological resource information system.
4. Maintain information on and evaluations of Commonwealth taxonomic collections.

Given the federal system of government in Australia, ABRS is a cooperative exercise aimed at coordinating, into a national framework, the major efforts made by state institutions. Particular ABRS objectives at present are to:

1. Promote the writing and publishing of *Flora of Australia.* This 50-60 volume work will be the first national Flora for over 100 years and the first to be written in Australia. The *Flora* is already widely recognized nationally and internationally as a significant reference work and a major aid to the identification of Australian flora. Two volumes per year are published now. Appendix 1 shows some sample text from published volumes.
2. Promote the writing and publishing of a 10-volume *Fauna of Australia,* to provide comprehensive information on the identification of Australia's fauna. Volume 1 (General Articles and Mammalia) is currently being written, with contributions from over 100 authors. This work complements the *Zoological Catalogue of Australia,* which will comprise approximately 70 volumes dealing with the taxonomy of Australian fauna at the species level. The *Zoological Catalogue* will also be accessible as a regularly updated online database. Appendix 2 shows sample text from published volumes dealing with vertebrates and invertebrates.
3. Develop a database system for distributional and taxonomic specimen data held by Australian museums and herbaria [known as the *Australian Biogeographic Information System* (ABIS)]. Occasional publications will be produced from this database. In preparation at present is an atlas of elapid (front-fanged) snakes in Australia and an atlas of Australian mangroves. Figure 1 shows a distribution map for the mangrove species *Aegiceras corniculatum.* This system also includes the mapping of Australia's vegetation.
4. Devise methodology to use ABIS in planning and co-ordinating biological surveys, environmental sampling, and land management.

Fig. 1. Distribution of *Aegiceras corniculatum* in Australia. The * indicates each site from which a specimen has been collected.

5. Improve and increase the output of taxonomic research and documentation in Australia by means of a Participatory Program. This Program calls for applications each year from taxonomic and biogeographic researchers. Each year certain "Preferred Objectives" are advertised, which relate to research, writing, and data compilation necessary to achieve the other ABRS objectives outlined above. A register of Australian and overseas workers interested in Australian biota is maintained, and directories are published from time to time. Statistics suggest that the taxonomic output, in terms of publications, has increased by 50% since the inception of ABRS.

DEVELOPMENT OF THE STUDY

The genesis of ABRS was in 1973, when an interim advisory body was appointed to report to the government by 1976 on the development of an Australian Biological Resources Survey.

In 1974, Dr. Franklyn Perring, then Director of the Biological Records Centre for the United Kingdom, made a study visit to Australia at the request of the ABRS Interim Council. After he had visited all Australian states and the mainland territories, a major symposium was held in Sydney. This symposium was attended by a large number of participants from the many regions of Australia. The symposium produced a major interchange of ideas and allowed Dr. Perring to present some ideas on biological survey and biological recording in Australia. In the years following that symposium, funding was provided for a wide range of biological survey-related activities, including taxonomic, biogeographic, and ecological studies. Ride (1978) presented a detailed review of the early years of the ABRS.

In 1978, the ABRS was formally established as an on-going activity, and the Bureau of Flora and Fauna was created to service the scientific and administrative needs of the Study. The organization of the ABRS and the Bureau is shown in the administration diagram (Figure 2).

In the last few years, project funding by ABRS has become strongly goal-directed, in contrast to prior attempts to fund proposals in all areas. This change was necessary to use available funding most effectively to achieve the *national* goals, rather than attempt to duplicate state efforts. Funding for taxonomic projects in the marine environment is also provided by the Marine Resources Allocation Advisory Council, which is working closely with ABRS to achieve the most efficient use of funds and expertise.

Major initiatives are now concentrating on the production of distribution atlases, the development and management of collecting strategies, and the implementation of ABIS at a national level. These activities include:

- ⋆ An atlas of the genus *Banksia*, coordinated and jointly funded by ABRS and the Western Australian Department of Conservation and Land Management. In addition to the expected contributions by professional biologists, this project is making widespread use of community effort to collect basic data. Use of community based efforts in biological survey has already been established by the *Atlas of Australian Birds* by Blakers, et al. (1984), and partly supported by ABRS.
- ⋆ A five-year plant collecting strategy, agreed to after discussion with the Council of Heads of Australian Herbaria. This program allows for ABRS funds to supplement state collecting efforts in a goal-directed fashion.

Since 1973, the ABRS has diverted its efforts from strategic funding of Biological Resource surveys to 1) tactical funding of specialized work to aid the publications program and 2) the development of ABIS. Interaction will be important between ABIS, as it develops, and the management of surveys by national and state bodies.

In all ABRS activities, full use is being made of advances in information technology to process, transfer, and allow access to the large volume of data accumulated in all aspects of the study.

RECENT DEVELOPMENTS

A significant development in this area has been the result of research program called BIOCLIM between the Bureau of Flora and Fauna and the CSIRO Division of Water and Land Resources. Briefly summarized, BIOCLIM produces climatic

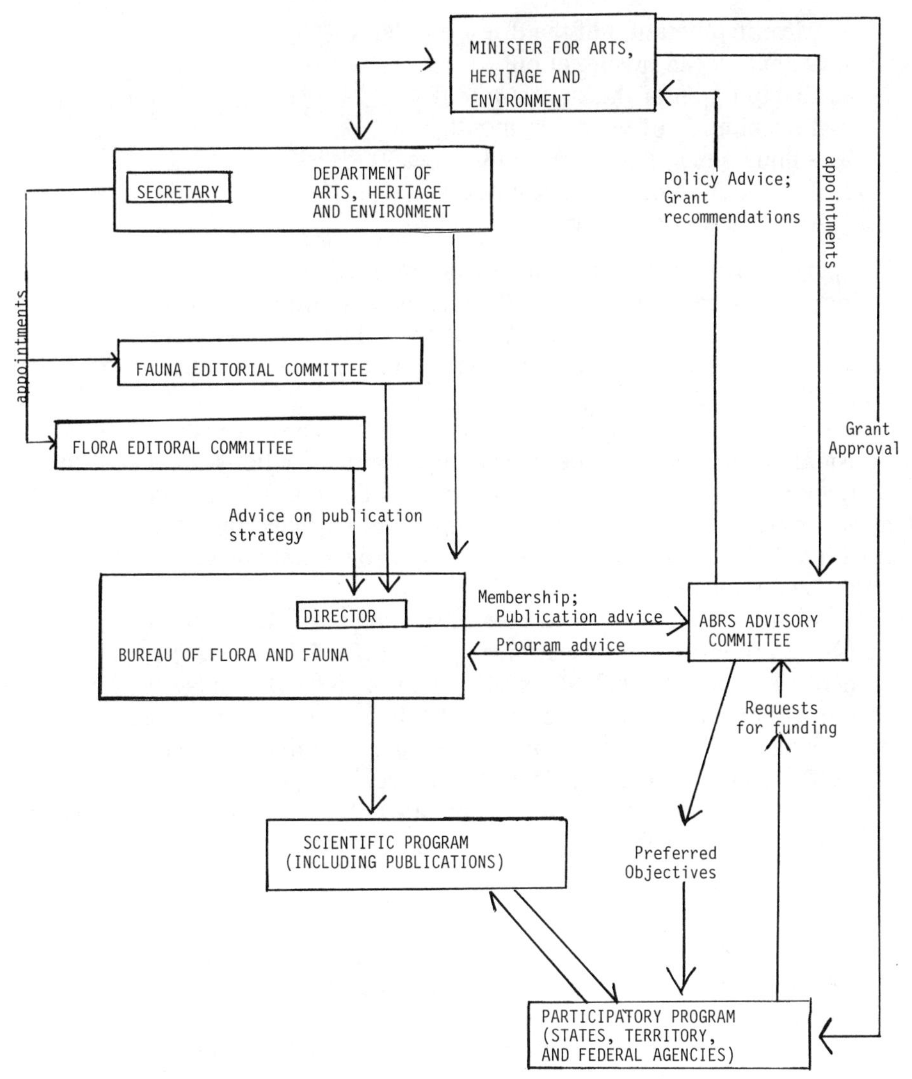

Fig. 2. The organization of the Australian Biological Resources Study (ABRS) and the Bureau of Flora and Fauna.

profiles for species based upon known points of occurrence. It then produces maps showing the known locations plus predicted locations. The climate profiles are based on a determination, for each known point, of 12 climatic parameters derived from monthly values for temperature and precipitation.

These parameters are as follows:

1. Annual mean temperature
2. Minimum temperature of the coldest month
3. Maximum temperature of the hottest month
4. Annual temperature range (3 minus 2)
5. Mean temperature of the wettest quarter (3 months)

6. Mean temperature of the driest quarter
7. Annual mean precipitation
8. Precipitation of the wettest month
9. Precipitation of the driest month
10. Annual precipitation range (8 minus 9)
11. Precipitation of the wettest quarter
12. Precipitation of the driest quarter

Temperatures are in °C, precipitations in mm.

The profiles are matched with the climate determined for each point of a 0.5 degree latitude-by-longitude grid of Australia. The degree of similarity in climate between the species profile and each grid point is noted on a five-point scale. All points are plotted on a map.

The predictive map is immediately useful. For example, the maps can assist decisions about whether apparently disjunct distributions are likely to reflect particular patterns or are merely artifacts of collecting. The maps can suggest new areas for field work, i.e. where a species may be expected to occur but has not yet been recorded. Predicted distributions may also suggest areas where a new species might be discovered.

These examples of suggested uses of the predicted distributions are but some that have arisen in the development of BIOCLIM. Taxa covered in the development of BIOCLIM included snakes, mammals, and a range of plant species. Figure 3 shows the actual and predicted distribution for a species of *Banksia*. Maps of all *Banksia* species in eastern Australia were distributed to all participants in the Banksia Atlas Project, described above.

Developments of this kind allow the data collected in a national survey to be utilized even more effectively to:

- Manage conservation reserves
- Assess the likely distribution of species thought to be endangered or rare
- Assess the likely possibilities of invasion by exotic species
- Select species for the rehabilitation of industrially or agriculturally degraded areas
- Develop strategies for further survey in remote locations

CONCLUSION

In answering the two questions posed at the start of this paper, ABRS not only accumulates scientific information but co-ordinates and sponsors tactical research. Such research aids in the resolution of environmental problems, and assists a wide range of organizations in planning appropriate environmental objectives.

In this context it is appropriate that ABRS is a component of the Department of Environment, rather than the Department of Science. For the future, ABRS has a major task to perform. Utilizing techniques offered by the rapid developments in information processing and communication will make that task simpler and allow greater access to the results of the study by Australians and all persons interested in Australian flora and fauna.

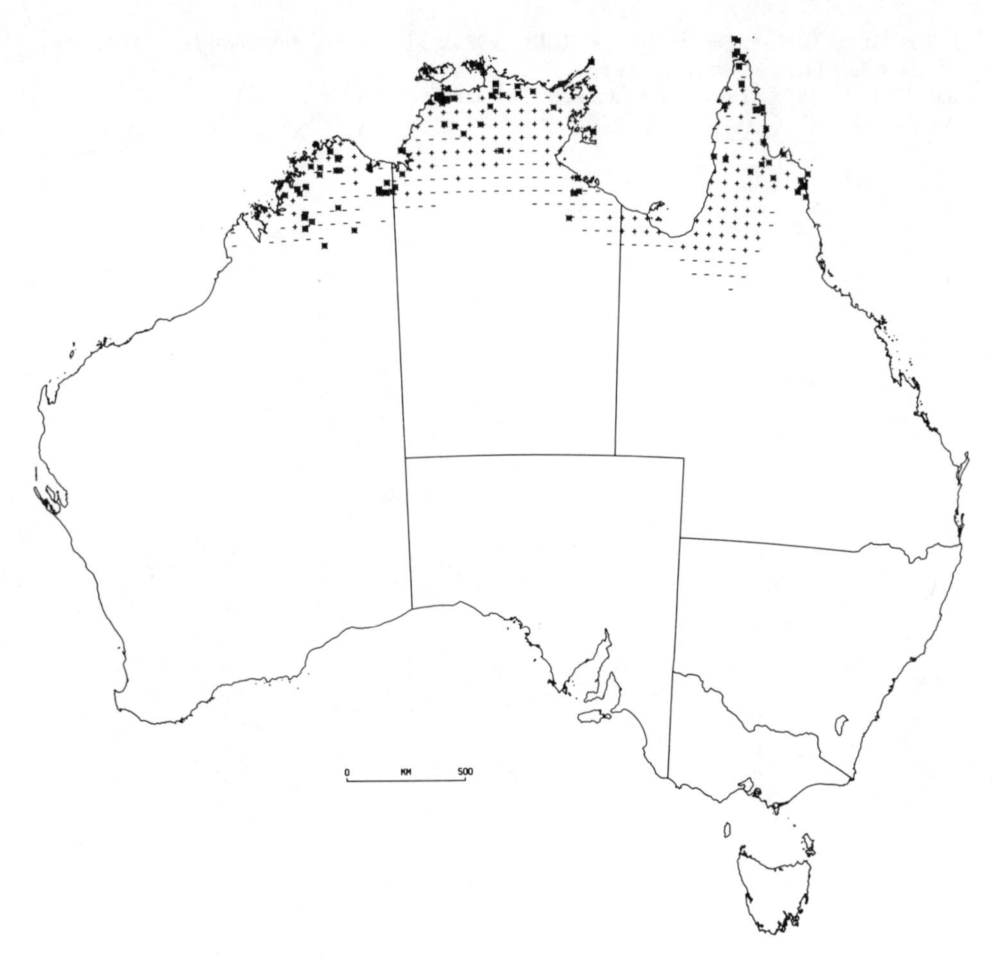

Fig. 3. Actual and predicted range of *Banksia dentata* in Australia. The * indicates each site from which a specimen has been collected, + indicates a predicted occurrence for sites within the 90 percentile climatic range and − indicates a predicted occurrence for sites between the 90 percentile and total climatic range.

ACKNOWLEDGEMENTS

I should like to acknowledge the efforts of all members of the ABRS Advisory and Editorial Committees during the last 7 years. Particular acknowledgement must go to Professor Sir Rutherford Robertson, Professor Ralph Slatyer, and Professor Geoffrey Sharman—past and present Chairmen of the ABRS Advisory Committee. Acknowledgement should also be given to previous Ministers for allowing the study to become established and to the Hon. Barry Cohen, Minister for Arts, Heritage and Environment for his support of the ongoing efforts of the ABRS. This paper represents the views of the author and not necessarily those of the Department of Arts, Heritage and Environment.

LITERATURE CITED

Anonymous 1984. *A national conservation strategy for Australia.* Proposed by a conference held in Canberra in June 1983. Australian Government Publishing Service, Canberra.

Blakers, M., S. J. J. F. Davies & P. N. Reilly. 1984. *The atlas of Australian birds.* Melbourne University Press, Melbourne. 738 pp.
Ride, W. D. L. 1978. Towards a National Biological Survey. *Search* 9: 73–82.

APPENDIX 1.

CHENOPODIACEAE

Paul G. Wilson

Herbs, shrubs, or (not in Australia) small trees, glabrous or pubescent, sometimes glandular. Leaves usually alternate, simple, often succulent, exstipulate, in the *Salicornieae* opposite and reduced to small lobes at the apex of jointed internodes (articles). Inflorescence of compact or open cymes or panicles, or reduced to solitary axillary flowers. Flowers small, monochlamydeous, bisexual or unisexual. Perianth of 1–5 tepals, often united, rarely absent, sometimes enlarged and developing wings, spines or tubercles in fruit. Stamens opposite and equal in number to perianth lobes or fewer, hypogynous or attached to wall of perianth; staminal disc present or absent; anthers exserted, bilocular, dehiscing by longitudinal slits. Ovary superior (half inferior in *Beta*) 2- or 3-carpellate, unilocular; stigmas usually 2 or 3. Ovule solitary, basal, campylotropous to amphitropous. Fruit a nut or berry with membranous, crustaceous, or succulent pericarp. Seed often lenticular; testa membranous to crustaceous; embryo straight, curved, horseshoe-shaped, annular, or spiral; albumen (perisperm) absent to abundant.

A cosmopolitan family of over 100 genera and 1500 species, particularly common in semi-arid environments and in saline habitats. Represented in Australia by 302 species in 28 native and 4 introduced genera. A few species have given rise to cultivars of agriculture; a number of the endemic species were important food plants of Australian Aborigines.

G. Bentham, Chenopodiaceae, *Fl. Austral.* 5: 150–208 (1870); F. Mueller, *Iconography of Australian Salsolaceous Plants*, Decades 1–9 (1889–91); E. Ulbrich, Chenopodiaceae, *Nat. Pflanzenfam.* 2nd edn, 16c: 379–584 (1934); P. Aellen, Chenopodiaceae, in Hegi, *Ill. Fl. Mitt.-Eur.* 2nd edn, 3: 534–747 (1960–61); R. C. Carolin *et al.*, Leaf structure in Chenopodiaceae, *Bot. Jahrb. Syst.* 95: 226–255 (1975); A. J. Scott, Reinstatement and revision of Salicorniaceae J. Agardh (Caryophyllales), *Bot. J. Linn. Soc.* 75: 357–374 (1978); A. J. Scott, A revision of the Camphorosmioideae (Chenopodiaceae), *Feddes Repert.* 89: 101–119 (1978).

In this treatment the text for 8 species of *Atriplex* has been contributed by G. A. Parr-Smith.

KEY TO TRIBES

1 Embryo curved to annular; albumen usually present

2 Plant with well-developed leaves; flowers not immersed in succulent spikes

3 Fruits not operculate; stigma papillose all over; ovary superior

4 Flowers usually in glomerules, axillary or paniculate; perianth not or little enlarged in fruit — **Trib. I. CHENOPODIEAE**

4: Flowers usually solitary and axillary; perianth usually enlarged, hardened, and bearing appendages at fruiting stage — **Trib. II. CAMPHOROSMEAE**

3: Fruits operculate; stigma papillose within; ovary semi-inferior — **Trib. III. BETEAE**

2: Plant leafless; stems jointed and succulent; flowers usually surrounded by succulent bracts — **Trib. IV. SALICORNIEAE**

1: Embryo spiral; albumen absent or scanty

12. ROYCEA

Roycea C. Gardner, *J. Roy. Soc. Western Australia* 32: 77 (1948); named after the Australian botanist R. D. Royce (1914–).

Type: *R. pycnophylloides* C. Gardner

Shrubs or perennial herbs, woolly or silky-pubescent when young. Leaves small, opposite or alternate, often fasciculate, entire, sessile, often spurred at base. Flowers inconspicuous, solitary, axillary, sessile, ebracteate, unisexual or bisexual. Perianth ovoid, c. 1 mm high, ±divided into 5 imbricate tepals, not enlarging in fruit. Stamens 5; filaments strap-shaped, united into a narrow disc at base. Ovary ovoid, c. 0.5 mm long, densely pubescent; style short, densely pubescent; stigmas 2 or 3, slender. Ovule erect, campylotropous; funicle arising from a cushion-like placenta. Fruit subglobular, 1–3 mm high, surrounded by perianth at base; pericarp thin, crustaceous. Seed horizontal or oblique; testa thin but slightly leathery; embryo circular surrounding a small central perisperm; radicle enclosed.

A genus of 3 species endemic in temperate and subtropical Western Australia.

1 Dense mat-forming herbaceous perennial with weak branches **1. R. pycnophylloides**

1: Erect shrub with rigid woody branches

2 Plant to 25 cm high, dioecious **2. R. spinescens**

2: Plant to 60 cm high; flowers bisexual **3. R. divaricata**

1. Roycea pycnophylloides C. Gardner, *J. Roy. Soc. Western Australia* 32: 78, t. II A–K (1948)

T: near Meckering, W.A., 7 Sept. 1945, *C. A. Gardner 7659*; holo: PERTH.

Illustrations: C. A. Gardner, *loc. cit.*

Perennial herb forming densely branched, silvery, mat-like growths to 1 m diam., dioecious. Branchlets closely woolly, obscured by the leaves. Leaves alternate, imbricate, narrowly triangular, naviculate and slightly cucullate at the acute apex, fleshy, c. 2 mm long, 1 mm wide, silky when young. Flowers towards apex of branches. Male flowers cup-shaped; tepals thin, ovate, c. 1 mm long, silky outside; anthers exserted; pistillode slender, c. 1 mm long, pubescent. Female flowers suborbicular c. 1 mm long; staminodes absent; stigmas long-exserted c. 4 mm long. Fruit broadly ovoid c. 2 mm high surrounded at base by persistent perianth; pericarp crustaceous. Fig. 37 I–J.

Endemic on the saline sandy flats around the Mortlock River near Meckering in southern W.A. Map 310.

W.A.: Meckering, *R. D. Royce 8413* (PERTH).

2. Roycea spinescens C. Gardner, *J. Roy. Soc. Western Australia* 32: 79, t. II L–S (1948)

T: near Meckering, W.A., 7 Sept. 1945, *C. A. Gardner 7659a*; holo: PERTH.

Illustrations: C. A. Gardner, *loc. cit.*

Small rigid shrub to 25 cm high forming colonies several metres in diameter, dioecious. Branches spinescent, divaricate, glabrous (pubescent in leaf axils). Leaves opposite (to alternate) in disjunct fascicles, ovate to triangular, 1–4 mm long, carinate, fleshy, glabrous, the larger ones spurred at base. Flowers in upper leaf-axils. Male flowers cup-shaped; tepals obovate, united in lower third, c. 2 mm long, ciliate; anthers exserted; pistillode slender, c. 1 mm long, pubescent. Female flowers globular; tepals sub-orbicular, ±free, c. 1 mm long, thin, ciliate; staminodes absent; ovary pubescent; stigmas c. 3 mm long. Fruit sub-globular, c. 3 mm high surrounded by persistent perianth; pericarp crustaceous. Fig. 37A–H.

Occurs from Morawa south to Merredin, W.A., in saline sand and sandy clay. Map 311.

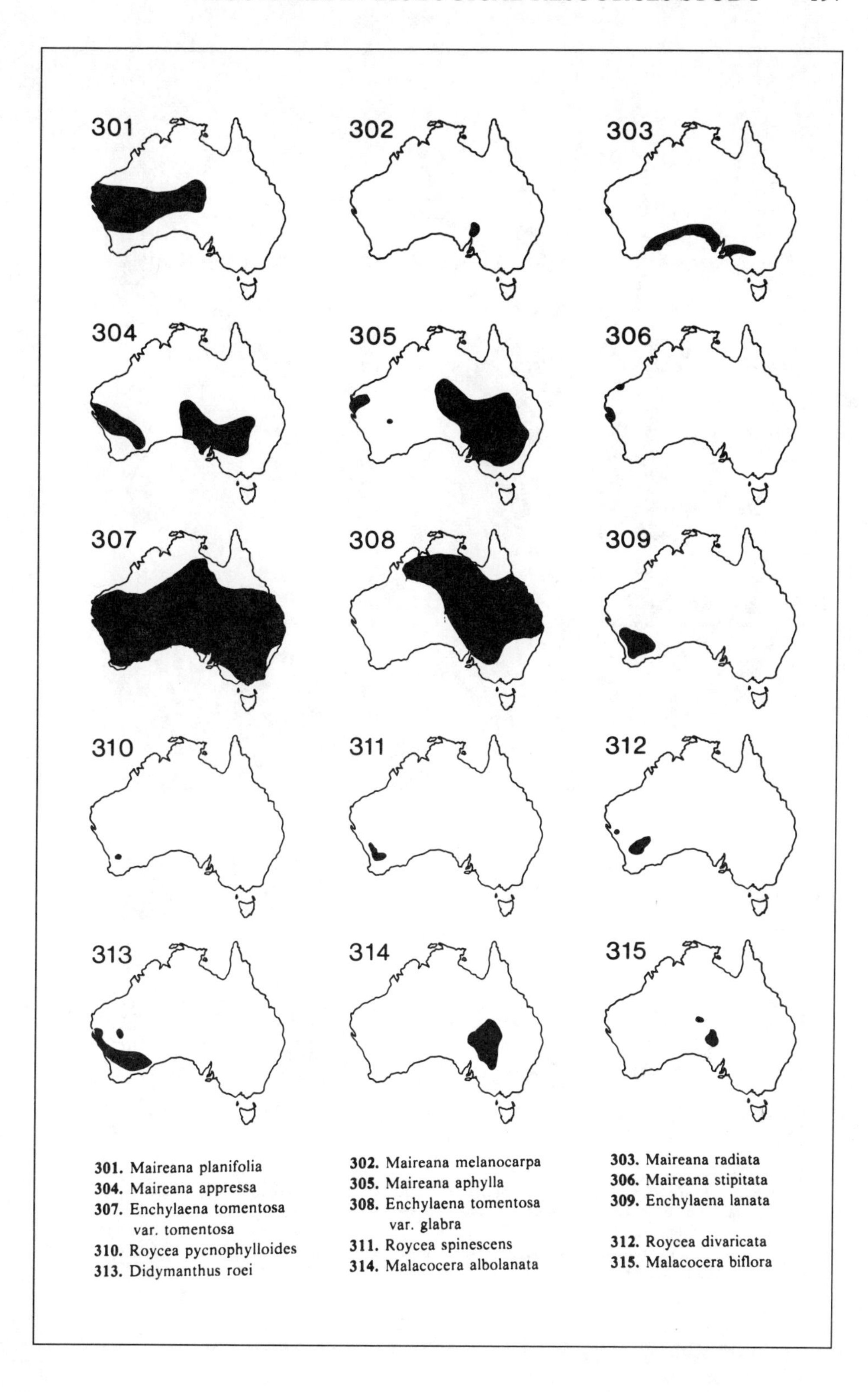

301. Maireana planifolia
304. Maireana appressa
307. Enchylaena tomentosa var. tomentosa
310. Roycea pycnophylloides
313. Didymanthus roei

302. Maireana melanocarpa
305. Maireana aphylla
308. Enchylaena tomentosa var. glabra
311. Roycea spinescens
314. Malacocera albolanata

303. Maireana radiata
306. Maireana stipitata
309. Enchylaena lanata
312. Roycea divaricata
315. Malacocera biflora

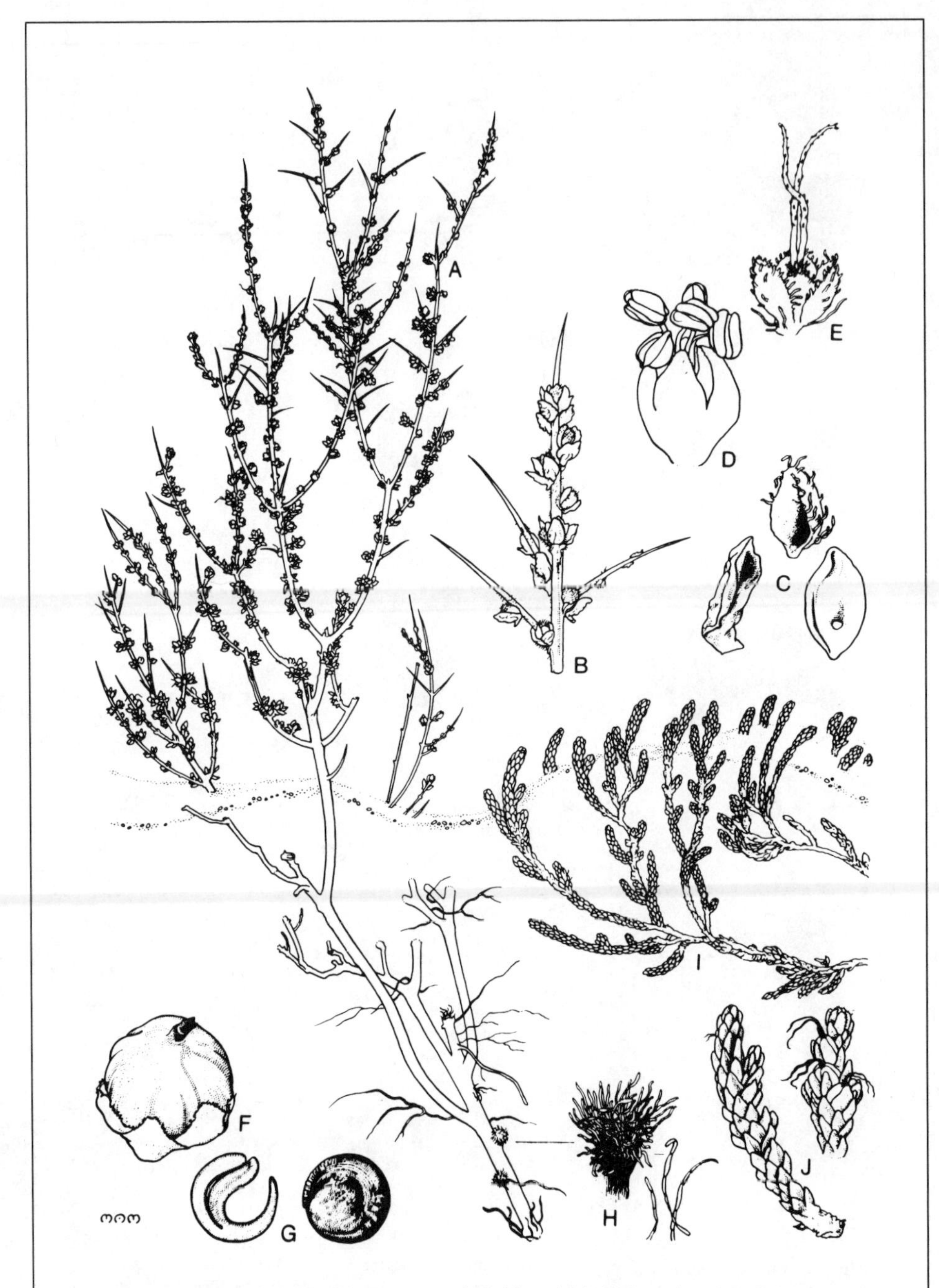

Figure 37. *Roycea.* **A–H**, *R. spinescens.* **A**, habit ×1.25; **B**, branch ×2.5; **C**, leaves ×5; **D**, male flower ×10; **E**, female flower ×10 (**A–E**, M. Menadue 36, PERTH). **F**, fruit ×10; **G**, embryo and seed ×10; **H**, hair cluster ×10 (**F–H**, P. Wilson 10964, PERTH). **I–J**, *R. pycnophylloides.* **I**, habit ×1.25; **J**, branch with female flowers ×2.5 (**I–J**, M. Menadue 37, PERTH).

APPENDIX 2

CARETTOCHELYDIDAE

INTRODUCTION

A cryptodirous family with a single living aquatic species which occurs in the rivers of southern New Guinea and Australia's Northern Territory.

Characterised in Australia by: bony shell covered by a fleshy, pitted skin without dermal scutes; paddle-shaped limbs, each with two claws; nostrils at the end of a prominent, fleshy proboscis.

References

Cann, J. (1978). *Tortoises of Australia.* Sydney : Angus & Robertson 79 pp.

Cogger, H.G. (1979). *Reptiles and Amphibians of Australia.* Sydney : A.H. & A.W. Reed 608 pp.

Pritchard, P.C.H. (1979). *Encyclopedia of Turtles.* Neptune, N.J. : T.F.H. Publications 895 pp.

Wermuth, H. & Mertens, R. (1961). *Schildkröten, Krokodile, Brückenechsen.* Jena: Gustav Fischer 422 pp.

Wermuth, H. & Mertens, R. (1977). Liste der rezenten Amphibien und Reptilien; Testudines, Crocodylia, Rhynchocephalia. *Das Tierreich* **100**: i–xxvii, 1–174

Carettochelys Ramsay, 1886

Carettochelys Ramsay, E.P. (1886). On a new genus and species of fresh water tortoise from the Fly River, New Guinea. *Proc. Linn. Soc. N.S.W. (2)* **1**: 158–162 [158] [1887 on title page of bound volume]. Type species *Carettochelys insculptus* Ramsay, 1886 by monotypy.

Carettocchelys Ramsay, E.P. (1886). On a new genus and species of fresh water tortoise from the Fly River, New Guinea. *Proc. Linn. Soc. N.S.W. (2)* **1**: 158–162 [158] [*errore pro Carettochelys* Ramsay, 1886; 1887 on title page of bound volume].

Carretochelys Ramsay, E.P. (1886). On a new genus and species of fresh water tortoise from the Fly River, New Guinea. *Proc. Linn. Soc. N.S.W. (2)* **1**: 158–162 [161] [*errore pro Carettochelys* Ramsay, 1886; 1887 on title page of bound volume].

Synonymy that of Cogger, H.G., this work.

This group is also found in southern New Guinea.

Carettochelys insculpta Ramsay, 1886

Carettocchelys [*sic*] ***insculptus*** Ramsay, E.P. (1886). On a new genus and species of fresh water tortoise from the Fly River, New Guinea. *Proc. Linn. Soc. N.S.W. (2)* **1**: 158–162 [158]. Type data: holotype, AM R3677, from Fly River, Papua New Guinea.

Carettochelys insculpta Boulenger, G.A. (1889). *Catalogue of the Chelonians, Rhynchocephalians, and Crocodiles in the British Museum (Natural History).* new edn. London : British Museum x 311 pp. [236] [valid emend. *pro Carettochelys insculptus* Ramsay, 1886].

Synonymy that of Cogger, H.G., this work.

Distribution: N coastal, (N Gulf), N.T.; extralimital in New Guinea. Ecology: lentic freshwater, lotic freshwater, aquatic, noctidiurnal, predator; seasonal breeder, oviparous, general carnivore. Biological references: Cogger, H.G. (1970). First record of the pitted-shelled turtle, *Carettochelys insculpta*, from Australia. *Search* **1**: 41; Schodde, R., Mason, I. & Wolfe, T.O. (1972). Further records of the pitted-shelled turtle (*Carettochelys insculpta*) from Australia. *Trans. R. Soc. S. Aust.* **96**: 115–117; Waite, E.R. (1905). The osteology of the New Guinea turtle (*Carettochelys insculpta*, Ramsay). *Rec. Aust. Mus.* **6**: 110–118.

SPHECIDAE

INTRODUCTION

This cosmopolitan family contains very small to very large solitary wasps. Australia has about 600 described species and subspecies in over 50 genera of which about a quarter are endemic. Adults are often collected on flowers or at nesting sites. Nests are made by burrowing in the ground, by using existing cavities in the ground, in dead wood or the pith of plants, by constructing mud cells in the open, on house walls or rocks or tree trunks, and by using abandoned mud nests. One genus (*Acanthostethus*) is cleptoparasitic. Adults of other genera provision their cells with insects - there are records from almost all the orders - or spiders or Collembola. Most genera exhibit some degree of prey specificity. *Bembix* is unusual in this respect, for while most northern hemisphere species studied prey on Diptera, about one third of the Australian species whose prey is known use other orders (Hymenoptera, Odonata and Neuroptera) and two species have been found to prey on more than one order of insects. Recent work on the biology of *Arpactophilus* sp., *Spilomena* sp. and *Pison* sp., not mentioned as a biological reference because the species were not identified, was published by Naumann (1983) and on *Lyroda* sp. by Evans & Hook (1984).

References

Bohart, R.M. & Menke, A.S. (1976). *Sphecid Wasps of the World : a Generic Revision.* Berkeley : Univ. California Press ix 695 pp.

Evans, H.E. & Hook, A.W. (1984). Nesting behaviour of a *Lyroda* predator (Hymenoptera : Sphecidae) on *Tridactylus* (Orthoptera : Tridactylidae). *Aust. Entomol. Mag.* **11**: 16–18

Evans, H.E. & West Eberhard, M.J. (1970). *The Wasps.* Ann Arbor : Univ. Michigan Press 265 pp.

Naumann, I.D. (1983). The biology of mud nesting Hymenoptera (and their associates) and Isoptera in rock shelters of the Kakadu Region, Northern Territory. *Aust. Natl. Parks & Wldlf. Serv. Spec. Publ. 10* pp. 127–189

Dolichurus Latreille, 1809

Dolichurus Latreille, P.A. (1809). *Genera Crustaceorum et Insectorum* secundem ordinem naturalem in familias disposita, iconibus exemplisque plurimis explicata. Paris : A. Koenig Vol. 4 397 pp. [387]. Type species *Pompilus corniculus* Spinola, 1808 by subsequent designation, see Latreille, P.A. (1810). *Considérations Générales sur l'Ordre Naturel des Animaux Composant les Classes des Crustacés, des Arachnides, et des Insectes;* avec un Tableau Méthodique de leurs Genres, Disposés en Familles. Paris : F. Schoell 444 pp. Compiled from secondary source: Bohart, R.M. & Menke, A.S. (1976). *Sphecid Wasps of the World : a Generic Revision.* Berkeley : Univ. California Press ix 695 pp.

This group is found worldwide, see Bohart, R.M. & Menke, A.S. (1976). *Sphecid Wasps of the World : a Generic Revision*. Berkeley : Univ. California Press ix 695 pp. [66].

Dolichurus carbonarius Smith, 1869

Dolichurus carbonarius Smith, F. (1869). Descriptions of new genera and species of exotic Hymenoptera. *Trans. Entomol. Soc. Lond.* **1869**: 301-311 [303]. Type data: holotype, BMNH *F. adult (seen 1929 by L.F. Graham), from Champion Bay, W.A.

Distribution: NE coastal, NW coastal, Qld., W.A.; only published localities Mackay, Kuranda, Dunk Is., Brisbane and Champion Bay. Ecology: larva - sedentary, predator : adult - volant; prey Blattodea, nest in pre-existing cavity. Biological references: Turner, R.E. (1915). Notes on fossorial Hymenoptera. XV. New Australian Crabronidae. *Ann. Mag. Nat. Hist. (8)* **15**: 62–96 (behaviour); Riek, E.F. (1955). Australian Ampulicidae (Hymenoptera : Sphecoidea). *Aust. J. Zool.* **3**: 131–144 (redescription).

Aphelotoma Westwood, 1841

Aphelotoma Westwood, J.O. (1841). *in*, Proceedings of the Entomological Society of London. (Descriptions of the following exotic hymenopterous insects belonging to the family Sphegidae). *Ann. Mag. Nat. Hist. (1)* **7**: 151–152 [152]. Type species *Aphelotoma tasmanica* Westwood, 1841 by monotypy.

Aphelotoma affinis Turner, 1910

Aphelotoma affinis Turner, R.E. (1910). Additions to our knowledge of the fossorial wasps of Australia. *Proc. Zool. Soc. Lond.* **1910**: 253–356 [341]. Type data: holotype, BMNH *F. adult (seen 1929 by L.F. Graham), from Townsville, Qld.

Distribution: NE coastal, Qld.; type locality only. Ecology: larva - sedentary, predator : adult - volant; prey Blattodea.

Aphelotoma auricula Riek, 1955

Aphelotoma auricula Riek, E.F. (1955). Australian Ampulicidae (Hymenoptera : Sphecoidea). *Aust. J. Zool.* **3**: 131–144 [139 pl 1 fig 8]. Type data: holotype, ANIC M. adult, from 10 mi S of Bowen, Qld.

Distribution: NE coastal, Qld.; only published localities near Bowen and Caloundra. Ecology: larva - sedentary, predator : adult - volant; prey Blattodea.

Aphelotoma fuscata Riek, 1955

Aphelotoma fuscata Riek, E.F. (1955). Australian Ampulicidae (Hymenoptera : Sphecoidea). *Aust. J. Zool.* **3**: 131–144 [139 pl 1 fig 7]. Type data: holotype, ANIC F. adult, from Catherine Hill, N.S.W.

Distribution: SE coastal, N.S.W.; type locality only. Ecology: larva - sedentary, predator : adult - volant; prey Blattodea.

Aphelotoma melanogaster Riek, 1955

Aphelotoma melanogaster Riek, E.F. (1955). Australian Ampulicidae (Hymenoptera : Sphecoidea). *Aust. J. Zool.* **3**: 131–144 [135 pl 1 figs 2–3]. Type data: holotype, ANIC M. adult, from Blundells, A.C.T.

Distribution: NE coastal, SE coastal, Murray-Darling basin, Qld., N.S.W., A.C.T. Ecology: larva - sedentary, predator : adult - volant; prey Blattodea.

Aphelotoma nigricula Riek, 1955

Aphelotoma nigricula Riek, E.F. (1955). Australian Ampulicidae (Hymenoptera : Sphecoidea). *Aust. J. Zool.* **3**: 131–144 [138 pl 1 fig 10]. Type data: holotype, ANIC M. adult, from Blundells, A.C.T.

Distribution: NE coastal, Murray-Darling basin, SE coastal, Qld., N.S.W., A.C.T.; only published localities Stanthorpe, Barrington Tops, Goulburn and Blundells. Ecology: larva - sedentary, predator : adult - volant; prey Blattodea.

Aphelotoma rufiventris Turner, 1914

Aphelotoma rufiventris Turner, R.E. (1914). New fossorial Hymenoptera from Australia and Tasmania. *Proc. Linn. Soc. N.S.W.* **38**: 608–623 [618]. Type data: holotype, BMNH *M. adult (seen 1929 by L.F. Graham), from Kuranda, Qld.

Distribution: NE coastal, Qld.; only published localities Kuranda, Bowen, Stradbroke Is., Caloundra and Stanthorpe. Ecology: larva - sedentary, predator : adult - volant; prey Blattodea. Biological references: Riek, E.F. (1955). Australian Ampulicidae (Hymenoptera : Sphecoidea). *Aust. J. Zool.* **3**: 131–144 (redescription).

Aphelotoma striaticollis Turner, 1910

Aphelotoma striaticollis Turner, R.E. (1910). Additions to our knowledge of the fossorial wasps of Australia. *Proc. Zool. Soc. Lond.* **1910**: 253–356 [341]. Type data: holotype, BMNH *F. adult (seen 1929 by L.F. Graham), from Townsville, Qld.

Distribution: NE coastal, Qld.; only published localities Townsville, Kuranda. Ecology: larva - sedentary, predator : adult - volant; prey Blattodea.

Aphelotoma tasmanica Westwood, 1841

Taxonomic decision of Riek, E.F. (1955). Australian Ampulicidae (Hymenoptera : Sphecoidea). *Aust. J. Zool.* **3**: 131–144 [136–137].

Aphelotoma tasmanica tasmanica Westwood, 1841

Aphelotoma tasmanica Westwood, J.O. (1841). *in* Proceedings of the Entomological Society of London. (Descriptions of the following exotic hymenopterous insects belonging to the family Sphegidae). *Ann. Mag. Nat. Hist. (1)* **7**: 151–152 [152]. Type data: syntypes (probable), OUM or BMNH *F. adult, from Tas.

Distribution: SE coastal, Vic., Tas. Ecology: larva - sedentary, predator : adult - volant; prey Blattodea.

Aphelotoma tasmanica auriventris Turner, 1907

Aphelotoma auriventris Turner, R.E. (1907). New species of Sphegidae from Australia. *Ann. Mag. Nat. Hist. (7)* **19**: 268–276 [269]. Type data: holotype, BMNH *M. adult, from Vic.

Biological Survey of Canada (Terrestrial Arthropods)

H. V. Danks
National Museum of Natural Sciences

Abstract: The Biological Survey of Canada, which currently comprises only the section on Terrestrial Arthropods, catalyses and coordinates studies in systematics and faunistics on behalf of the National Museum of Natural Sciences and the Entomological Society of Canada. The Survey consists of a small Secretariat and a larger advisory committee. The Survey acts as a clearing house for information and also synthesizes scientific information, initiates and coordinates specific scientific projects of particular current importance, and prepares commentaries on matters of national concern. It has been successful in helping to characterize the fauna because it is steered by the scientific community, produces material of scientific value, has a small and efficient central operation, and considers ecological as well as taxonomic aspects of the fauna.
Keywords: Clearing-house, Entomology, Systematics, Information Service.

INTRODUCTION

The Biological Survey of Canada was developed to help characterize the Canadian fauna by supporting appropriate work based on a national scientific overview of requirements. Because the Survey developed relatively recently from an initiative of the Entomological Society of Canada (see below), only terrestrial arthropods are currently included. The Survey organization comprises a small full-time Secretariat, employed at the National Museum of Natural Sciences, and a larger, widely representative advisory Scientific Committee constituted through the Entomological Society of Canada. The full Committee meets twice per year, and its expenses are provided by the National Museum of Natural Sciences.

ROLES OF THE BIOLOGICAL SURVEY OF CANADA

The Survey helps to expand knowledge of the Canadian fauna by catalysing and coordinating faunal studies. It therefore does not, for example, maintain a collection, nor will it attain great size, because it aims mainly to facilitate work carried out in various federal, provincial, university, and other institutions. The Survey functions both as a clearing house and in scientific capacities.

The Role of the Survey as a Clearing House for Information

The Survey maintains current directories and inventories of personnel with interests in systematics and faunistics, collections, sites with long-term protection suitable for faunal work, field stations, and other more specialized inventories. The major inventories are published from time to time (e.g., Biological Survey Project, 1977, 1978). From this information, the Survey can answer queries and assist those persons or agencies planning fieldwork in a particular area. The head of the Secretariat travels extensively each year to entomological centers across the country to exchange information and ideas, so that entomologists with overlapping interests in different organizations can be put in touch with one another and can contribute to the scientific discussions of the Survey. The role of information exchange is also incorporated into a twice-yearly *Newsletter of the Biological Survey of Canada (Terrestrial Arthropods)*, which includes each spring a list of requests for cooperation or information.

Scientific Roles of the Survey

In its main role, the Survey, with its advisory Committee, oversees national scientific endeavours in systematic and faunistic entomology. Three sorts of activities support and stimulate basic research toward an understanding of the fauna: syntheses of knowledge, the discussion and organization of individual scientific projects, and more general initiatives.

1. Syntheses of Knowledge

The Survey plans and executes substantial reviews of scientific information. Major volumes already published are *Canada and its Insect Fauna* (Danks, 1979), *Arctic Arthropods* (Danks, 1981), and *Temporal and Spatial Changes in the Canadian Insect Fauna* (Downes, 1981). Other books and symposia proceedings covering broad fields of study are in preparation (e.g., see Yukon below).

Canada and its Insect Fauna established a baseline for the arthropod survey. It explored the physical environment of Canada and its history, the habitats and arthropod distributions that have resulted, the state of knowledge for each group, and general problems concerning the nature of the fauna in relation to Canadian conditions. *Arctic Arthropods* took a similar approach in greater depth for areas beyond the northern limit of trees. Taxonomy and ecology were treated at more-or-less equal length in these books, reflecting the philosophy that a scientifically appropriate survey cannot be planned from only one of these perspectives.

2. Scientific Projects

Most of the current activities catalysed by the Survey are focused into specific scientific projects, selected for their current scientific importance and feasibility. Most of the projects deal with subjects, taxa, or regions that are especially significant to an understanding of the Canadian insect fauna and that are so broad in scope that unaided efforts would be piecemeal or not feasible. Other projects develop subject areas that have previously been overlooked (see also soil fauna in the next section). Examples of some of these projects, at varying stages of development, are outlined below to indicate how diverse topics can be approached in this way.

-*Illustrated keys to the families of arthropods in Canada.* Many laboratory and field studies, and teaching, are seriously hindered by the lack of an up-to-date and readily usable key to the families of terrestrial arthropods found or expected to be found in Canada. A profusely illustrated key to families is therefore being developed. A fascicle on myriapods is complete and awaits publication, and work on the fascicles covering insects is proceeding.

-*Arthropod fauna of the Yukon.* The Yukon territory has many diverse habitats and a very extensive but inadequately understood arthropod fauna. For this reason, and because part of the region was unglaciated during Pleistocene time, the Yukon is a key area for interpreting the nature and development of the Canadian fauna. Biological Survey initiatives have led to several recent field parties, and a preliminary prospectus has been developed for a book on the arthropod fauna of the Yukon.

-*Arthropod fauna of Canadian grasslands.* The arthropods of the grasslands are surprisingly inadequately known. This hinders the understanding of the fauna of a large part of the continent and in particular an understanding of the origin and setting of the faunas of major present-day agricultural lands. An annual *Grasslands Newsletter* is being produced as this project develops.

-*Insect fauna of freshwater wetlands in Canada.* Wetlands cover a substantial area of the surface of Canada and are of particular ecological importance as reservoirs that buffer the effects of variations in rainfall on the lands that surround them. The insect fauna of wetlands, the identification of larvae (the stage normally encountered), and their ecological characteristics are largely unknown. Peatlands (bogs and fens) comprise the greatest proportion of wetlands in Canada; marshes are of particular interest because they serve as breeding and staging grounds for a wide variety of birds, many of which depend on arthropods as food. After this project had been introduced to the entomological community (Rosenberg, 1981), a symposium was organized (1984) in which participants reviewed the status of each characteristic group of wetland insects in bogs, fens, and marshes. The proceedings of this symposium are now being edited prior to publication.

-*Aquatic insects of Newfoundland.* The fauna of Newfoundland is of particular interest in a Canadian context because it presumably reflects mainly postglacial immigration from the mainland to an island or peninsular situation. Aquatic habitats cover one third of the surface area of the island of Newfoundland, yet their insects were very inadequately known when the project on this fauna began. A preliminary survey of insect groups—the first stage of the project—has now been completed (Larson and Colbo, 1983), revealing a characteristically boreal but much reduced fauna.

-*Arthropod fauna of freshwater springs in Canada.* Springs are discrete habitats that can be relatively easily sampled. Their faunas would be expected to provide an index of groundwater quality, and springs are particularly interesting zoogeographically because they may contain endemic species and can indicate the presence or absence of recent glaciation. This project was launched with a preliminary article (Williams, 1983) and is being supported by individual research contributions, by preparation of a bibliography (awaiting publication), and by other activities. A symposium to synthesize available information is planned for the near future.

3. General Initiatives

The Survey prepares commentaries to alert individuals or organizations and to stimulate studies in areas of concern. A few examples of these activites are outlined below.

-*Arthropods of the soil.* The Survey recognized the great importance of the very inadequately known soil fauna of Canada in maintaining the fertility of soils but realized that the taxonomic resources available to remedy deficiencies in knowledge were too limited to support an active project. Therefore, the ecological roles of the arthropod fauna of the soil and the current state of knowledge of Canadian soil arthropods were outlined in a brief (Marshall et al., 1982). Various activities including wide distribution of this brief helped to stimulate an international conference on faunal influences on soil structure (1984), the proceedings of which are in press. This conference established contact between pedologists and soil zoologists and has led to plans for further relevant research.

-*Appraisal of environmental disturbance.* The Survey prepared a brief (Lehmkuhl et al., 1984) which pointed out that insects are potentially valuable in assessing environmental disturbance and noted that such work has to be planned and conducted with proper scientific procedures.

-*Regional collections.* The Survey realized the need for a network of regional collections in addition to a strong national collection and has presented ideas about the use and development of these collections for discussion by interested biologists (Danks, 1983).

DEVELOPMENT OF THE BIOLOGICAL SURVEY OF CANADA

Origin of the Biological Survey of Canada

The Biological Survey began as the Pilot Study for a Biological Survey of the Insects of Canada, an idea proposed by the Entomological Society of Canada in 1974 (Entomological Society of Canada, 1974). This idea stemmed from the realization that the insect fauna of Canada, despite its scientific and practical importance, was very inadequately known: about 66,000 species of terrestrial arthropods are believed to occur, but only about half of these have even been described. An understanding of the fauna of Canada was therefore by no means commensurate with needs relating not only to insects of agricultural, forestry, and medical significance, but also to the more diffuse requirements of environmental concerns and the management of complex, living natural resources. Moreover, systematic resources were inadequate for the tasks required, especially in the absence of any group charged with a national scientific overview of research and needs in insect taxonomy and ecology.

The Pilot Study was funded by a government contract held by the Entomological Society of Canada. Results of this study and of subsequent more specific contracts showed that a survey of the form already described was feasible. Following the recommendations of the Pilot Study (see Danks, 1978), a Survey organization was established in 1980 within the National Museum of Natural Sciences and was renamed the Biological Survey of Canada (Terrestrial Arthropods). The important role of the Entomological Society of Canada was retained, that of appointing the advisory committee, which provides access to the scientific community. The form of the name accords with the idea that the terrestrial arthropods module of the Survey is a model for parallel arrangements in due course for other groups of organisms.

Future of the Biological Survey of Canada

The Survey recently has prepared (for the National Museum of Natural Sciences) documents confirming that the Survey should be expanded by the addition of modules for other areas of study but without the dilution of the disciplinary expertise of the existing module. Each module would thus rely on the scientific community for a particular discipline, through an advisory Scientific Committee, and would have its own small Secretariat. All such modules would be coordinated, and a complete Biological Survey of Canada would be composed of relatively few modules, each of rather broad range. Further expansion of the Survey therefore requires two parallel developments: interest from other groups of scientists, and additional federal resources channelled through the National Museum of Natural Sciences.

CONCLUDING REMARKS

The Biological Survey of Canada differs in three main characteristics from some existing biological surveys in other countries (such surveys were reviewed by Danks and Kosztarab, in press). First, the Biological Survey of Canada draws its ideas and scientific orientations from the scientific community itself (by direct contact and through the advisory Scientific Committee), rather than from a narrowly appointed executive group.

Second, it relies largely on the coordination of widespread existing resources by a small and efficient organization, without the creation of a large new central research group. However, in various ways (e.g., publication of briefs, development of ideas and proposals) the Survey provides a climate favorable for obtaining new resources to support work that is viewed as important. Moreover, the Survey's coordination results in usable published scientific products as explained above, so that the Survey does not simply produce administrative documents for "coordination".

Third, the Survey does not restrict its purview to taxonomic work. Because the fauna can be characterized and information about it used in various ways only if the functioning of species as well as their identity is known, the Survey has always emphasized ecology as well as taxonomy. This approach has been successful in initiating and maintaining the Biological Survey of Canada (Terrestrial Arthropods) during a period of widespread reduction in governmental resources.

LITERATURE CITED

Biological Survey Project. 1977. *Annotated list of workers on systematics and faunistics of Canadian insects and certain related groups.* Pilot Study for a Biological Survey of the Insects of Canada, Entomological Society of Canada, Ottawa, Ontario. 107 p.

Biological Survey Project. 1978. Collections of Canadian insects and certain related groups. Pilot study for a biological survey of the insects of Canada, Entomological Society of Canada. *Bull. Entomol. Soc. Can.* 10(1), Suppl., 21 p.

Danks, H. V. 1978. Biological survey of the insects of Canada. *Bull. Entomol. Soc. Can.* 10(3): 70–73.

Danks, H. V.(ed.) 1979. Canada and its insect fauna. *Mem. Entomol. Soc. Can.* 108. 573 p.

Danks, H. V. 1981. *Arctic arthropods. A review of systematics and ecology with particular reference to the North American fauna.* Entomol. Soc. of Canada, Ottawa. 608 p.

Danks, H. V. 1983. Regional collections and the concept of regional centres. *In*: D. J. Faber (ed.) Proceedings of 1981 Workshop on Care and Maintenance of Natural History Collections. *Syllogeus* 44: 151–160. 196 p.

Danks, H. V. & M. Kosztarab. In press. Biological surveys. (*Proceedings of a Symposium on Biosystematics Services in Entomology*, Hamburg, 1984).

Downes, J. A. (ed.) 1981. Symposium: temporal and spatial changes in the Canadian insect fauna. *Can. Entomol.* 112(1980)(11): 1089–1238.

Entomological Society of Canada. 1974. A biological survey of the insects of Canada: a brief. *Bull. Entomol. Soc. Can.* 6(2), Suppl., 16 p.

Larson, D. J. & M. H. Colbo. 1983. The aquatic insects: biogeographic considerations. *In*: G. R. South (Ed.), *Ecology and biogeography of the island of Newfoundland.* Monographiae Biologicae, Vol. 48. Junk, The Hague.

Lehmkuhl, D. M., H. V. Danks, V. M. Behan-Pelletier, D. J. Larson, D. M. Rosenberg, & I. M. Smith. 1984. Recommendations for the appraisal of environmental disturbance: some general guidelines, and the value and feasibility of insect studies: a brief by the Biological Survey of Canada (Terrestrial Arthropods). *Bull. Entomol. Soc. Can.* 16(3), Suppl., 8 p.

Marshall, V. G., D. K. McE. Kevan, J. V. Matthews, Jr., & A. D. Tomlin. 1982. Status and research needs of Canadian soil arthropods: a brief by the Biological Survey of Canada (Terrestrial Arthropods). *Bull. Entomol. Soc. Can.* 14(1), Suppl., 5 p.

Rosenberg, D. M. 1981. Aquatic insects of freshwater wetlands. *Bull. Entomol. Soc. Can.* 13(4): 151–153.

Williams, D. D. 1983. National survey of freshwater springs. *Bull. Entomol. Soc. Can.* 15(1): 30–34.

SECTION VI.

CONCLUSION

An Overview of the Symposium

Lorin I. Nevling, Jr.
Field Museum of Natural History
(President, ASC 1984-1985)

Abstract: An overview is presented of the 1985 ASC symposium, "Community Hearings on a National Biological Survey," based on the presentations, comments, questions, and discussion. The overview focuses on the basic points of general agreement that can serve as a basis for a continuing dialogue and organizational effort.

The 1984 meeting of the Association of Systematics Collections held in Champaign/Urbana, Illinois can be described in a single word—tumultuous. The matter that triggered the stormy session was the multiplicity of viewpoints and honest disagreements concerning the national biological survey.

It became increasingly clear during the course of the 1984 meeting that the survey was of such significance to the scientific community and to the public at large that it should be the focus of the Association's 1985 Symposium. Drs. K. C. Kim and Lloyd Knutson volunteered to organize the symposium, an act which seemed scarcely sane at the time. As they began to develop the form of the symposium and to develop the list of potential contributors who would provide the substance, their organizational skills became evident. As you reflect on this symposium, I am certain that you will gain an appreciation of the care with which it was organized and the plain hard work it entailed. They did a superb job, and we are grateful to them for organizing this opportunity for all of us.

This symposium outlined the rationale for a national biological survey. Included were discussions on the usefulness and applicability of survey results to a wide variety of federal agencies, the scientific community, and other important segments of the private or governmental sector.

Technical details of data management, manipulation, and dissemination were presented, together with discussions of data pitfalls and shortages. Technological breakthroughs in hardware and software in the relatively near future were introduced; this topic left some of us in management positions wondering if it would be prudent to purchase anything at this time.

The past and current responsibilities held by various governmental groups demonstrated the bio-political complexity of undertaking the survey and brought to mind the motto on the flag of Culpeper's Minutemen, "Don't Tread on Me." We also heard of the plans, activities, and results of similar projects from our colleagues in Australia, the Republic of Mexico, and subsequently, in this volume, from Canada. They have struggled with many of the same questions that have faced us during the course of the symposium.

Overall, I sensed that an increasingly intellectual approach to the survey has replaced the emotionalism of last year's ASC meeting and that a conceptual model is beginning to evolve.

The rationale—the "case"—for the survey is emerging but, in my opinion, is still short of being fully convincing. The survey must not be over-sold as a social program for the biological community or as a matter of nationalistic pride. The case for relevancy must be established on merit and must be convincing to legislators who need to see a tangible benefit to their constituents and to the nation. I believe I detected general agreement by the participants on the following points:

⋆ The activities of man are destroying the natural world.
⋆ The national biological survey would be useful to a broad constituency including decision makers, and informed decisions may help to retard the rate of destruction of our natural heritage.
⋆ If a national biological survey were to be mounted, a substantial financial participation by the federal government would be required.
⋆ Finances will drive programs and determine their scope.
⋆ The survey will require both basic research and synthesis of previous research. An active interface between and among scholars and information specialists will need to be established.
⋆ A massive amount of information is available concerning our biological resources, but it is scattered and sometimes inaccessible. The chain leading from information to knowledge to understanding needs to be understood and strongly forged from the outset.
⋆ The survey must be focused, and that focus must be *simple* and *understandable*.
⋆ Short-term results will be required and progress must proceed in an incremental, logical fashion. The products and markets must be carefully and realistically determined in advance.
⋆ A management framework needs to be developed to formulate broad goals and specific objectives for the survey.

We have met together, shared mutual opportunities, renewed old acquaintances, established new contacts, and come to realize the broad base of support the survey has from many diverse quarters. This is the first of many planning and organizational sessions, and I am pleased that the Association of Systematics Collections was the organizer. My thanks to Drs. Kim and Knutson, to the speakers, to all the participants, and to Dr. Craig Black, who invited us to hold our meeting here at the Natural History Museum of Los Angeles County.

SUMMARY OF RECOMMENDATIONS PRESENTED BY SPEAKERS[1]

Robert M. West
Carnegie Museum of Natural History

W. Donald Duckworth
Bernice P.Bishop Museum

1. Institutionalize a national biological survey at state and federal levels, in close coordination with adjoining states and countries; consult extensively with existing agencies and organizations already or potentially performing the same work; conduct workshops and symposia to ensure adequate understanding and assignment of tasks.
2. Establish a central archive and data maintenance facility to develop or adapt consistent data-gathering processes and networks; this central archive will include life-history data, photographs, molecular genetic information, and so on.
3. Conduct a general survey and inventory of the status of existing biotic resources, including endangered or fragile ecosystems. The boundaries of this survey should be biologically, not geopolitically, determined.
4. Resist the temptation to establish a national biological survey on an insufficient funding basis. Review the successes and failures of other countries' efforts via a specially constituted steering committee. Simultaneously, assess the actual present and projected needs of potential user communities.
5. Provide an effective means of distribution of and access to all the data gathered and generated by this survey, taking advantage of current and future technology and the changing and variable requirements of the user communities. Include keys and identification manuals in addition to monographic studies.

[1]The body of recommendations presented in the separate papers was an important product of the conference. The conference co-chairmen thus felt it would be instructive to ask two of the participants to attempt to summarize the essence of the recommendations and pertinent discussions.

Epilogue

A National Biological Survey: A Vehicle for the Study of Our Living Resources

The concept of a national biological survey is a critical issue for environmental protection, resource management, and systematic biology in the U.S. It provides a powerful vehicle for "rediscovering" the state of the North American fauna and flora, through which the essential database on the living resources of our nation will be developed. Before engaging in environmental debates, such as the impact of acid rain and its importance to our environment or planning our resource management, we first must have an accurate assessment of the North American biota and its rapidly changing status. This concept is expected to promote a national consensus to develop suitable national programs to explore, document, and monitor our fauna and flora. However, many questions will require precise answers before any national consensus can be reached pertaining to the approaches and organization for an appropriate national biological survey and its relationship to other national endeavors.

The 1985 ASC Symposium was the first national conference in the U.S. specifically devoted to a biological survey of this country. This Symposium dealt to some extent with questions of funding, organizational basis, structure, and functioning, which were beyond the initial goals of the meeting. However, it is hoped that the Symposium will provide a keystone for further debates on many aspects of a national biological survey so that a reasonably national consensus can be reached on the approaches and organization for such a national program. We expect that a few more national conferences on a national biological survey will be held. We trust that this volume will be of value as baseline information for future discussion.

Ke Chung Kim
The Pennsylvania State University

Lloyd Knutson
Biosystematics and
Beneficial Insects Institute